Subfactors
and Knots

Conference Board of the Mathematical Sciences

CBMS

Regional Conference Series in Mathematics

Number 80

Subfactors and Knots

Vaughan F. R. Jones

Published for the
Conference Board of the Mathematical Sciences
by the
American Mathematical Society
Providence, Rhode Island
with support from the
National Science Foundation

Expository Lectures
from the CBMS Regional Conference
held at the US Naval Academy
June 5–11, 1988

This research was partially supported by National Science Foundation Grant DMS-8714130.

1991 *Mathematics Subject Classification*. Primary 46L10, 57M25 ; Secondary 16S34, 22E65, 22E67, 22E70, 46L37, 46L40, 57M15, 81T30, 82B23.

Library of Congress Cataloging-in-Publication Data
Jones, Vaughan F. R., 1952–
  Subfactors and knots/Vaughan F. R. Jones.
  p. cm. —(Regional conference series in mathematics: no. 80)
  Includes bibliographical references.
  ISBN 0-8218-0729-3
  1. Von Neumann algebras—Congresses. 2. Knot theory—Congresses.
I. Conference Board of the Mathematical Sciences. II. Title. III. Series.
QA1.R33 no. 80
[QA326]
510 s—dc20   91-24438
[512′.55]   CIP

Copying and reprinting. Individual readers of this publication, and nonprofit libraries acting for them, are permitted to make fair use of the material, such as to copy an article for use in teaching or research. Permission is granted to quote brief passages from this publication in reviews, provided the customary acknowledgment of the source is given.

Republication, systematic copying, or multiple reproduction of any material in this publication (including abstracts) is permitted only under license from the American Mathematical Society. Requests for such permission should be addressed to the Manager of Editorial Services, American Mathematical Society, P.O. Box 6248, Providence, Rhode Island 02940-6248.

The owner consents to copying beyond that permitted by Sections 107 or 108 of the U.S. Copyright Law, provided that a fee of $1.00 plus $.25 per page for each copy be paid directly to the Copyright Clearance Center, Inc., 27 Congress Street, Salem, Massachusetts 01970. When paying this fee please use the code 0160-7642/91 to refer to this publication. This consent does not extend to other kinds of copying, such as copying for general distribution, for advertising or promotional purposes, for creating new collective works, or for resale.

Copyright ©1991 by the American Mathematical Society. All rights reserved.
Printed in the United States of America
The American Mathematical Society retains all rights
except those granted to the United States Government.
The paper used in this book is acid-free and falls within the guidelines
established to ensure permanence and durability. ∞
This publication was typeset using $\mathcal{A}_\mathcal{M}\mathcal{S}$-TEX,
the American Mathematical Society's TEX macro system.

10 9 8 7 6 5 4 3 2   96 95 94 93 92 91

Contents

Introduction	ix
Lecture 1. von Neumann Algebras	1
1.1. Three topologies on $\mathscr{B}(\mathscr{H})$	1
1.2. von Neumann's bicommutant theorem	2
1.3. (Concrete) von Neumann algebras	2
1.4. Factors	3
1.5. Examples of factors	4
1.6. Comparison of projections	6
1.7. Types I, II_1, II_∞, and III factors	7
1.8. Standard form for II_1 factors	7
1.9. The fundamental group of a II_1 factor	8
1.10. Type III factors	9
1.11. Hyperfiniteness	10
Lecture 2. Group Actions and Subfactors	13
2.1. The coupling constant	13
2.2. Galois theory for finite group actions	14
2.3. Connes' results on automorphisms of R	15
2.4. Extensions of Connes' automorphism results	16

2.5. Index for subfactors	17
2.6. The basic construction	18
2.7. The basic construction in finite dimensions	19
2.8. Two basic constructions, proof of Goldman's theorem	20

Lecture 3. Values of the Index, Virasoro Algebra — 23

3.1. Values of the index	23
3.2. The Virasoro unitarity result	24
3.3. The continuous series for subfactors	25
3.4. Iterating the basic construction: the e_i algebra	26
3.5. Combinatorics of the e_i's	26
3.6. The e_i algebra is a II_1 factor	27
3.7. The element $e_1 \vee e_2 \vee \cdots \vee e_n$, the values $4\cos^2 \frac{\pi}{n}$	28
3.8. Ghosts	30

Lecture 4. Construction of Examples, Further Structure — 33

4.1. The discrete series of subfactors	33
4.2. Bratteli diagrams of the e_i algebras	36
4.3. Affine Lie algebra	38
4.4. Realizing the Virasoro discrete series	39
4.5. The tower of relative commutants	40
4.6. Examples of towers of relative commutants	42
4.7. The relative commutant problem	43

Lecture 5. The Braid Group and Its Representations — 45

5.1. Definition and presentation	45
5.2. Action of the braid group on the free group	47
5.3. The pure braid group and the inductive structure of the braid groups	47

5.4. Burau and Gassner representations	48
5.5. Representations in the e_i algebras	50
5.6. Representations in the Pimsner-Popa-Temperley-Lieb algebra (PPTL)	52
5.7. QISM representations of the braid group	53
5.8. The Potts model and Gaussian representations	55
5.9. More representations	56

Lecture 6. Knots and Links — 59

6.1. Knots and links	59
6.2. The fundamental group and the Alexander module	60
6.3. Seifert surfaces	62
6.4. Seifert matrices, S-equivalence	63
6.5. Untwisted doubles of knots have trivial Alexander module	65
6.6. Skein relation for the Alexander polynomial	66
6.7. Closed braids and the Burau representation	67

Lecture 7. The Knot Polynomial V_L — 69

7.1. First definition of V_L	69
7.2. The theory of plats	70
7.3. A second definition of V_L, the plat approach	71
7.4. Kauffman's e_i diagrammatics	72
7.5. Skein relation, third definition of V_L	73
7.6. The skein polynomial, inductive definition	74
7.7. The Kauffman polynomial	75
7.8. Kauffman's "states model", fourth and best definition of V_L	76

Lecture 8. Knots and Statistical Mechanics — 79

8.1. Statistical mechanics formalism	79

8.2. Ising, Potts, Vertex, Spin, and IRF models 79

8.3. Transfer matrices 82

8.4. The six-vertex model, Temperley-Lieb equivalence 83

8.5. Commuting transfer matrices, the Yang-Baxter equation 85

8.6. Vertex models on link diagrams 87

8.7. Spin models on link diagrams 89

Lecture 9. The Algebraic Approach — 93

9.1. The Hecke algebra 93

9.2. The relationship between the e_i algebra and $H(q,n)$ 94

9.3. Ocneanu's trace on $H(q,n)$ 95

9.4. Positivity considerations and subfactors from the Hecke algebra — 96

9.5. The Birman-Murakami-Wenzl algebra 98

9.6. The Markov trace on the BMW algebra 100

9.7. Structure of the BMW algebra 101

9.8. Wenzl's result on Brauer's centralizer algebra 102

9.9. Quantum invariant theory 103

Appendix 105

References 107

Introduction

These are the notes of a CBMS series of lectures I gave at Annapolis in the spring of 1988. The lectures were addressed to an audience consisting of low-dimensional topologists and operator algebraists. I tried to make the material comprehensible for both groups. This means that there is an extensive introduction to the theory of von Neumann algebras, and another to knot theory and the braid groups. The material presented in these notes is more or less exactly what was covered in the lectures. One exception is the definition of the knot polynomial $V(t)$. In the lectures I began with Kauffman's bracket as a definition and in the notes I end with it. Thus the notes are ordered historically in this respect.

It was a pleasure to give lectures where both knot theory and von Neumann algebras were treated, as well as some elementary material from statistical mechanics and conformal field theory. Since the spring of 1988 the whole area has undergone tremendous development, most notably in terms of the deepening connections with physics. Witten's topological quantum field theory and his invariants for three manifolds have been the most visible part of this work. It was tempting to rewrite the notes to incorporate some of the new developments, but I decided to leave them exactly as they were after the lectures, only adding occasional footnotes with indications of subsequent progress.

Thus some parts of the text seem a little naive, for instance the veiled implication that index for subfactors and central charge of Virasoro representations are directly related. Much progress on these connections has been made by Wassermann.

The choice of topics was, of course, highly personal. Thus the reader will not find much on the detailed classification of subfactors. This is also because the situation was still somewhat unclear in 1988, there being no available proofs of the main results.

I would like to thank G. Price, M. Kidwell, B. Baker and all others responsible for organizing this CBMS series.

Vaughan Jones
August 1991

Lecture 1. von Neumann Algebras

1.1. Three topologies on $\mathscr{B}(\mathscr{H})$. If $\mathscr{H}$ is a complex Hilbert space with inner product $\langle\ ,\ \rangle$, the *norm* topology on the *-algebra $\mathscr{B}(\mathscr{H})$ of all bounded linear operators on $\mathscr{H}$ is defined by the norm: $\|x\| = \sup_{\substack{\xi \in \mathscr{H} \\ \|\xi\| \leq 1}} \|x\xi\|$.

In finite dimensions x is a matrix and $\|x\|^2$ may be calculated as the largest eigenvalue of x^*x. The same is true in infinite dimensions if we replace "largest eigenvalue" by "spectral radius", where the spectral radius of an operator a is $\sup\{|\lambda| : \lambda - a \text{ is not invertible}\}$. If we consider $\mathscr{H} = L^2([0, 1], dx)$, $L^\infty([0, 1], dx)$ acts on $\mathscr{H}$ by pointwise multiplication and the norm of $f \in L^\infty$ is the essential sup of $|f|$. Thus the continuous functions $C([0, 1])$ form a norm closed subalgebra of $L^\infty([0, 1], dx)$ on $\mathscr{H}$. (Note that the choice of $([0, 1], dx)$ is inessential. The same things are true for any compact space and measure you are likely to think of in the next ten minutes.)

The *strong* topology on $\mathscr{B}(\mathscr{H})$ is that defined by the seminorms $x \mapsto \|x\xi\|$ as ξ runs through $\mathscr{H}$. Thus a sequence (or net if you must) x_n converges to x iff $x_n\xi$ converges to $x\xi$ in $\mathscr{H}$ for all $\xi \in \mathscr{H}$. The strong topology is much weaker than the norm topology. In fact we will soon see that, for the example of $L^\infty([0, 1])$ and $C([0, 1])$ acting on $L^2([0, 1])$, $C([0, 1])$ is actually strongly dense in $L^\infty([0, 1])$. To see a sequence that converges strongly without converging in norm, let x_n be the characteristic function of $[0, 1/n]$ viewed as an element of L^∞. Obviously x_n tends strongly to zero, but $\|x_n\| = 1$ for all n.

The *weak* topology on $\mathscr{B}(\mathscr{H})$ is that defined by the seminorms $x \mapsto |\langle x\xi, \eta \rangle|$ as ξ and η run through $\mathscr{H}$. The Cauchy–Schwarz inequality shows that the strong topology is stronger than the weak topology. In fact the weak topology is so weak that the unit ball of $\mathscr{B}(\mathscr{H})$ is weakly compact—which is often very useful. Probably the simplest example of a sequence of operators tending weakly but not strongly to zero is the sequence $e^{in\theta}$ in $L^\infty(S^1)$ (on $L^2(S^1)$), which by Fourier series is the same as the obvious shift operator on $\ell^2(\mathbb{Z})$.

1

The interplay between the above three topologies is basic to von Neumann algebras. We refrain from mentioning the many other topologies around.

1.2. von Neumann's bicommutant theorem. Let us prove the following simplified version of von Neumann's bicommutant theorem (see [vN1]). We use the following standard notation: if $S \subseteq \mathscr{B}(\mathscr{H})$ then $S' = \{x \in \mathscr{B}(\mathscr{H}) : xs = sx \text{ for all } s \in S\}$, and $S'' = (S')'$.

THEOREM. *Let S be a subset of $\mathscr{B}(\mathscr{H})$ with the following two properties*:
  a) *If $x \in S$ then $x^* \in S$*;
  b) $1 \in S$ (1 *is the identity operator on $\mathscr{B}(\mathscr{H})$*).

Then $\mathrm{alg}(S)$ is strongly (hence weakly) dense in S''. ($\mathrm{alg}(S)$ *is the algebra generated by S.*)

PROOF. First check that $\mathrm{alg}(S) \subseteq S''$. Now suppose $y \in S''$. What we must show is this: for any finite set $\xi_1, \ldots, \xi_n$ in $\mathscr{H}$, there is an element x of $\mathrm{alg}(S)$ with $x\xi_i$ arbitrarily close to $y\xi_i$ for all i. Let us suppose at first that we only want to approximate one vector $y\xi$. The trick is this: let V be the closure of the vector subspace $\mathrm{alg}(S)\xi$ and let p be the operator that is orthogonal projection onto V. Clearly $aV \subseteq V$ for all $a \in S$; so by property a), $ap = pa$. Thus $yp = py$ since $y \in S''$. So $yV \subseteq V$. But by property b), $\xi \in V$ so that $y\xi \in \overline{\mathrm{alg}(S)\xi}$, which is precisely what we wanted to prove.

The general case of $\xi_1, \ldots, \xi_n$ involves another trick which is used all over the subject: make $\xi_1, \xi_2, \ldots, \xi_n$ into a single vector on the Hilbert space $\bigoplus_{i=1}^{n} \mathscr{H}$. Then $\mathrm{alg}(S_i)$ and y act diagonally on $\bigoplus_{i=1}^{n} \mathscr{H}$ and we can, after making some matrix calculations to see how commutants behave under this "amplification" of $\mathscr{H}$, repeat the previous argument with ξ replaced by $\bigoplus_{i=1}^{n} \xi_i$ to conclude the proof. □

NOTE. If 1 did not belong to S the theorem still applies provided one cuts down to the closed subspace of $\mathscr{H}$ that is all that S notices.

This beautiful little theorem shows that two notions, one analytic (closure in the strong topology) and one purely algebraic (being equal to one's bicommutant) are the same for *-subalgebras of $\mathscr{B}(\mathscr{H})$ containing 1. It thoroughly justifies the definition of §1.3. Note also that the theorem shows that "strongly closed" and "weakly closed" are the same thing for a *-subalgebra of $\mathscr{B}(\mathscr{H})$.

1.3. (Concrete) von Neumann algebras.
DEFINITION. If $\mathscr{H}$ is a complex Hilbert space, a *von Neumann algebra* is a *-subalgebra M of $\mathscr{B}(\mathscr{H})$ containing 1 such that either M is strongly (weakly) closed or $M = M''$. If S is a selfadjoint subset of $\mathscr{B}(\mathscr{H})$ then S'' is the von Neumann algebra generated by S.

EXAMPLES.
 i) The algebra $\mathscr{B}(\mathscr{H})$ itself is certainly closed, thus a von Neumann algebra.
 ii) The algebra $L^\infty([0,1], dx)$ is easily shown to be its own commutant, thus a von Neumann algebra.
 iii) If G is a group and $g \mapsto u_g$ is a unitary representation of G, then the commutant $\{u_g\}'$ is a von Neumann algebra.
 iv) If $\mathscr{H}$ is finite-dimensional, it is not too hard to see that a von Neumann algebra M is just a direct sum of matrix algebras corresponding to some orthogonal decomposition $\mathscr{H} = \mathscr{H}_1 \oplus \cdots \oplus \mathscr{H}_k$ so that the matrices in M will look like

$$\dim \mathscr{H}_1 \left\{ \overbrace{\begin{bmatrix} x_1 & & \\ & \ddots & \\ & & x_1 \end{bmatrix}}^{\dim \mathscr{H}_1} \oplus \begin{bmatrix} x_2 & & \\ & \ddots & \\ & & x_2 \end{bmatrix} \oplus \cdots \oplus \overbrace{\begin{bmatrix} x_k & & \\ & \ddots & \\ & & x_k \end{bmatrix}}^{\dim \mathscr{H}_k} \right\} \dim \mathscr{H}_k$$

where the x_i's are matrices.
 v) If M on $\mathscr{H}$ and N on $\mathscr{K}$ are von Neumann algebras there are obvious notions of direct sum $M \oplus N$ on $\mathscr{H} \oplus \mathscr{K}$ and tensor product $M \otimes N$ on $\mathscr{H} \otimes \mathscr{K}$.

We list some important facts about von Neumann algebras.
 1) The set of all projections of a von Neumann algebra M forms a complete (orthomodular) lattice. M is generated by its projections since it contains the spectral projections of any selfadjoint element.
 2) Abelian von Neumann algebras are completely understood. As well as example ii) above there is $\ell^\infty(\mathbb{N})$ on $\ell^2(\mathbb{N})$ and obvious reductions and combinations with example ii). One must allow some kind of "multiplicity" as can be seen in finite dimensions. But on a separable Hilbert space that is the whole story. Probably the best way to deal with the multiplicity question is to relegate it to the spectral theorem and state, as von Neumann did, the structure theorem for abelian von Neumann algebras as the fact that they are generated by a single selfadjoint operator.
 3) von Neumann algebras can be abstractly characterized as C^*-algebras which are duals as Banach spaces. See [Sa].

1.4. Factors. The center $Z(M)$ of a von Neumann algebra is abelian. So by fact 2 of §1.3 we know everything about it. In finite dimensions it would be a direct sum of copies of $\mathbb{C}$, one for each summand in the decomposition of example 4 of §1.3. In general, using the spectral theory, von Neumann defined ([vN2]) (in the separable situation) a notion of "direct integral" of Hilbert spaces $\int_X^\oplus \mathscr{H}(\lambda) d\mu(\lambda)$ so that, for instance, for

$L^\infty([0, 1])$ on $L^2([0, 1])$ the corresponding decomposition of $L^2([0, 1])$ would be $\int_{[0,1]}^\oplus \mathscr{H}(\lambda)\, dx(\lambda)$ where $\mathscr{H}(\lambda) \equiv \mathbb{C}$. The whole algebra M respects this decomposition and we end up with a notion of direct integral of von Neumann algebras: $M = \int_X^\oplus M(\lambda)\, d\lambda$ on $\int_X^\oplus \mathscr{H}(\lambda)\, d\lambda$, the whole decomposition being essentially unique. The individual $M(\lambda)$'s will have trivial center (to get a feel for this, work it out in finite dimensions). Thus any von Neumann algebra is the direct integral of ones with trivial center.

Although the technical details of this theory are rather messy, and it can usually be avoided by "global" methods, the direct integral decomposition is tremendously helpful in trying to visualize a von Neumann algebra on a basic level. Of course one does not get any further than the $M(\lambda)$'s with trivial center.

DEFINITION. A von Neumann algebra M whose center is just the scalar multiples of the identity is called a *factor*.

EXAMPLES.

a) $\mathscr{B}(\mathscr{H})$ is a factor.
b) In finite dimensions a factor will always be of the form $\mathscr{B}(\mathscr{H}) \otimes \mathbb{C}\,\mathrm{id}$ on $\mathscr{H} \otimes \mathscr{K}$. This is also true in infinite dimensions provided the factor is isomorphic, as an abstract algebra, to some $\mathscr{B}(\mathscr{H})$.

Example b) explains the name "factor"—such factors correspond to tensor product factorizations of the Hilbert space. The remarkable fact, discovered by Murray and von Neumann in their works [**MvN1, 2, 3**] is that not all factors are like this, and indeed, as we shall see, it is not very difficult to construct examples.

1.5. Examples of factors. a) Let Γ be a discrete group (e.g., the free group on two generators) all of whose conjugacy classes are infinite, except that of the identity (we will call such groups i.c.c.). Let $\gamma \to u_\gamma$ denote the left-regular representation of Γ on $l^2(\Gamma)$. As matrices on $l^2(\Gamma)$ with respect to the obvious basis indexed by $\gamma \in \Gamma$, the u_γ, and hence all elements of $\mathrm{alg}(\{u_g\})$ are of the form $x_{\gamma,\nu} = f(\gamma^{-1}\nu)$ (forgive me if the inverse is in the wrong place) for some function f of finite support on Γ. The same is true for weak limits of such operators except that f will no longer be of finite support. However, applying the operator to the basis element for the identity we see that f is in ℓ^2. It is thus convenient and accurate to write elements of $M = \{u_\gamma\}''$ as sums $\sum_{\gamma \in \Gamma} f(\gamma) u_\gamma$ where $f \in \ell^2$ (although not all l^2 functions define elements of M). The sense of convergence of the sum will be clear later on. In any case, in order that $\sum_{\gamma \in \Gamma} f(\gamma) u_\gamma$ belong to the center of M, it must commute with u_ν for all ν, which implies $f(\nu\gamma\nu^{-1}) = f(\gamma)$, i.e., f is constant on conjugacy classes. But f is in ℓ^2 and all nontrivial conjugacy classes are infinite. Thus the support of f is the identity so that M is a factor. Call it vN(Γ).

One may see quickly that this factor is not as in example b) of §1.4 by observing that the linear function $\text{tr}(\sum f(\gamma)u_\gamma) = f(\text{identity})$ has the property $\text{tr}(ab) = \text{tr}(ba)$ and is not identically zero. It is simple to show that no such function exists on $\mathscr{B}(\mathscr{H})$ unless $\dim \mathscr{H} < \infty$.

b) The previous example was an example of a very general construction called the *crossed product*, where one begins with a von Neumann algebra N on $\mathscr{H}$ and a group Γ acting by automorphisms on N (in example a), $N = \mathbb{C}$) and one forms a von Neumann algebra $M = N \rtimes \Gamma$ (on $\mathscr{H} \otimes \ell^2(\Gamma)$) generated by $u_\gamma = \text{id} \otimes u_\gamma$ and an action of N on $\mathscr{H} \otimes \ell^2(\Gamma)$. All elements of $N \rtimes \Gamma$ can be represented as sums $\sum_{\gamma \in \Gamma} x_\gamma u_\gamma$, $x_\gamma \in N$, and $u_\gamma x u_\gamma^{-1} = \gamma(x)$ (the action of γ on x) for $x \in N$. It is then trivial to show that the following conditions together suffice to imply that $N \rtimes \Gamma$ is a factor.

(i) The action of Γ is "free", i.e., $xy = y\gamma(x)$ for all $x \in N$ implies $y = 0$ or $\gamma = 1$.

(ii) The algebra of fixed points for Γ is a factor.

Crossed products may also be formed by continuous (locally compact) groups, but they are algebraically less transparent.

c) Let us give an important example of the previous construction. The group will be $\mathbb{Z}$ and N will be $L^\infty(S^1)$. The generator of $\mathbb{Z}$ will act by an irrational rotation. As in example a) there is a trace functional on $L^\infty(S^1) \rtimes \mathbb{Z}$ given on $\sum_{n \in \mathbb{Z}} f_n u_n$ by $\int_{S^1} f_0(\theta)\, d\theta$. This example can obviously be varied by replacing $\mathbb{Z}$ by any discrete group and S^1 by any finite measure space, provided the group action preserves the measure and is free and ergodic. It was recognized very early on that in this situation the crossed product algebra depends only on the *equivalence relation* defined on the measure space by the orbits of the group action, indeed that it is possible to define the crossed product algebra given only the measure space and the equivalence relation (with countable equivalence classes). For the definitive treatment see [**FM**].

d) *The G.N.S. construction* provides an elementary but useful way to pass from a *-algebra which is not necessarily complete to a von Neumann algebra. The necessary data are a *-algebra A and $\varphi : A \to \mathbb{C}$ with $\varphi(a^*a) \geq 0$. One then forms a Hilbert space by defining a not necessarily definite inner product on A by $\langle a, b \rangle = \varphi(b^*a)$. The Hilbert space $\mathscr{H}_\varphi$ is then the completion of the quotient of A by the kernel of this form. Under favorable circumstances (such as if A is a C^*-algebra), A will act on $\mathscr{H}_\varphi$ by left multiplication. This representation of A is called the G.N.S. representation. The von Neumann algebra generated by the image of A in this representation should be thought of as a *completion* of A with respect to φ. In general, it is difficult to say if the G.N.S completion is a factor or not. One often meets surprises where A has trivial center but its completion does not.

To illustrate the procedure let us give an alternative way of constructing the example 1.5 a). On the group algebra $\mathbb{C}\Gamma$ of finite sums $\sum_\gamma c_\gamma \gamma$ one defines $\gamma^* = \gamma^{-1}$ and $\operatorname{tr}(\sum_\gamma c_\gamma \gamma) = c_{\mathrm{id}}$. It is clear that the G.N.S. Hilbert space $\mathscr{H}_{\mathrm{tr}}$ is naturally the same as $l^2(\Gamma)$ and the G.N.S. representation of $\mathbb{C}\Gamma$ on it is just the linear extension of the left-regular representation.

Of more interest, especially for these lectures, are the following examples. Let (A_n, φ_n) be an increasing union of finite-dimensional von Neumann algebras together with compatible linear functionals φ_n (i.e., $\varphi_{n+1}|_{A_n} = \varphi_n$). Then the union or inductive limit of the A_n's is a *-algebra and the φ_n's define a φ on it for which the G.N.S. construction works perfectly. One obtains many interesting factors in this way. The simplest nontrivial example occurs when $A_n = \bigotimes_{i=1}^n M_2(\mathbb{C})$, the inclusion of A_n in A_{n+1} is a $\hookrightarrow \begin{pmatrix} a & 0 \\ 0 & a \end{pmatrix}$ and φ_n is the trace, normalized so that $\varphi_n(\mathrm{id}) = 1$. Once again the G.N.S. completion of A admits a trace so is not $\mathscr{B}(\mathscr{H})$. We will have occasion to examine many more examples in the course of these lectures.

1.6. Comparison of projections.

DEFINITION. If M is a von Neumann algebra on $\mathscr{H}$ and p and q are projections in M, we say that $p \precsim q$ if there is an operator $u \in M$ with $uu^* = p$ and $u^*uq = u^*u$ (or equivalently, u^*u is a projection onto a subspace of $\mathscr{H}$ contained in $q\mathscr{H}$, written $u^*u \leq q$). We say that p and q are equivalent ($p \sim q$) if there is u in M with $uu^* = p$, $u^*u = q$. It is true that $p \sim q$ if $p \precsim q$ and $q \precsim p$ (see [MvN1]).

The point is that the operator u must be in M so that the notion of comparison depends heavily on M. If $M = \mathscr{B}(\mathscr{H})$ it is trivial that two projections are equivalent if and only if their images have the same dimension. Thus the idea was born that equivalence classes of projections represent an abstract notion of dimension for an arbitrary factor. The first result confirming this is the following.

THEOREM. *If M is a factor and p, q are projections in M then either $p \precsim q$ or $q \precsim p$.*

A proof may be found in [MvN1]. The result and the proof are quite natural if one considers the analogy with ergodic theory, p and q corresponding to measurable subsets of a measure space on which a group is acting. Ergodicity corresponds to being a factor and $p \precsim q$ means that the set p admits a (countable) partition into subsets p_i for each of which there is an element g_i of the group with $g_i(p_i) \subseteq q$ and $g_i(p_i) \cap g_j(p_j) = \varnothing$ for $i \neq j$. In fact I have heard tell, but never followed up the references, that the whole theory of comparison of projections, including the type I, II, III classification of §1.7, was done in the ergodic theory context by E. Hopf in [Hop], before [MvN1].

1.7. Types I, II$_1$, II$_\infty$, and III factors. So far the only way we have distinguished between factors was by the existence or otherwise of a trace function. The comparison of projections will give a related but more precise tool.

We begin by calling a projection q *finite* if $p \leq q$, $p \sim q$ implies $p = q$ and *infinite* if there is a $p \sim q$ with $p \lneq q$. A projection $p \neq 0$ is called *minimal* if it dominates no other projection in M other than 0.

DEFINITION. A factor M is of type I, II$_1$, II$_\infty$, or III according to the following mutually exclusive conditions:

type I: M has a minimal projection;

type II$_1$: M has no minimal projections and every projection is finite;

type II$_\infty$: M has no minimal projections but it has both finite and infinite projections;

type III: M has no finite projections except 0.

It is easy to see that if M has a trace tr with $\text{tr}(x^*x) > 0$ for $x \neq 0$ then it is finite-dimensional or of type II$_1$. It is also fairly easy to prove that if M is of type I it is like $\mathscr{B}(\mathscr{H}) \otimes \text{id}$ on $\mathscr{H} \otimes \mathscr{K}$, and that any type II$_\infty$ factor is a tensor project of a II$_1$ and a type I factor.

Murray and von Neumann showed in [**MvN1**] that if M is a factor there is an essentially unique "dimension function" d: projections of $M \to [0, \infty]$ subject to

(i) $d(0) = 0$,
(ii) $d(\sum_{i=1}^{\infty} p_i) = \sum_{i=1}^{\infty} d(p_i)$ if $p_i \perp p_j$ for $i \neq j$;
(iii) $d(p) = d(q)$ if $p \sim q$.

It follows that $d(p) = d(q) \Rightarrow p \sim q$ and that d may always be normalized so that its range is as follows:

type I: $\{0, 1, 2, \ldots, n\}$ with $n = \infty$ possible;

type II$_1$: $[0, 1] =$ the whole unit interval;

type II$_\infty$: $[0, \infty]$;

type III: $\{0, 1\}$.

It should be clear at this stage that examples 1.5 a) and c) are both type II$_1$ factors. One can "see" the dimension function on projections in example 1.5 c) by examining the abelian subalgebra $L^\infty(S^1)$. A projection in here is the characteristic function of some set and its dimension is its (normalized) Haar measure. Thus continuous dimensionality is not a mysterious phenomenon at all.

In these examples the dimension function comes from a trace and in [**MvN2**] it is shown that any II$_1$ factor has a unique normalized trace extending its dimension function.

1.8. Standard form for II$_1$ factors. A II$_1$ factor M, considered as an abstract complex *-algebra, possesses a (unique) trace tr which is a state of the kind for which the G.N.S. construction may be performed. The resulting Hilbert space completion of M for the inner product $\langle a, b \rangle = \operatorname{tr}(b^*a)$ is denoted $L^2(M, \operatorname{tr})$ or often $L^2(M)$. The reason for this notation is that if M were $L^\infty(X, \mu)$ for some nice probability space (X, μ) and if $\operatorname{tr}(f)$ were $\int_X f \, d\mu$ then $L^2(M)$ would be the Hilbert space $L^2(X, \mu)$. Indeed, there is a highly developed theory of "noncommutative L^p spaces" for II_1 factors, where the p-norm $\|x\|_p$ of $x \in M$ is defined to be $(\operatorname{tr}(|x|^p))^{\frac{1}{p}}$; see [Di2], [N].

The continuity properties of tr are such that the algebra of operators on $L^2(M)$ defined by left multiplication by M is already weakly closed so that M acts on $L^2(M)$ as a von Neumann algebra. This action is called the *standard form* of M. Since $\operatorname{tr}(ab) = \operatorname{tr}(ba)$, right multiplication by elements of M also extends to give bounded operators on $L^2(M)$ to give an action of the opposite von Neumann algebra M^{opp} on $L^2(M)$. The situation is completely symmetric and $M' = M^{\text{opp}}$. It is important to see that the symmetry between the left and right operations of M is implemented by a conjugate linear isometry $J : L^2(M) \to L^2(M)$, which is simply the extension to $L^2(M)$ of the map $x \to x^*$ defined on M. It is a trivial calculation that, if $\xi \in L^2(M)$ then $\xi x = Jx^*J\xi$ so that $JMJ = M'$. For full details see [Di1].

One must think of the standard form as being nothing but the left regular representation of M.

1.9. The fundamental group of a II_1 factor. The continuous dimensionality of II_1 factors makes them look somewhat homogeneous. If q is a nonzero projection in a II_1 factor M it follows from basic theory that qMq is also a II_1 factor, which one might guess to be isomorphic to M. Further thought shows that there is no good reason for thinking this, though notice that this isomorphism property depends only on the trace (hence equivalence class) of q. The fundamental group encodes the set of all traces of projections q for which $M \cong qMq$.

The best way to define the fundamental group of M is to consider the type II_∞ factor $M \otimes \mathscr{B}(\mathscr{H})$ of infinite matrices over M. This $M \otimes \mathscr{B}(\mathscr{H})$ has an infinite trace tr and an automorphism α of $M \otimes \mathscr{B}(\mathscr{H})$ may multiply tr by a positive real constant λ. (There are no minimal projections to normalize tr by and $\operatorname{tr}(1) = \infty$.) The set $\{\lambda | \lambda \in \mathbb{R}$ and there is $\alpha \in \operatorname{Aut}(M \otimes \mathscr{B}(\mathscr{H}))$ with $\operatorname{tr} \circ \alpha = \lambda \operatorname{tr}\}$ is obviously a group and is called the fundamental group of M.

We see that if $\lambda < 1$ is in the fundamental group and p is a minimal projection of $\mathscr{B}(\mathscr{H})$ then $(1 \otimes p)M \otimes \mathscr{B}(\mathscr{H})(1 \otimes p) \cong M$, and if α is an automorphism with $\operatorname{tr} \circ \alpha = \lambda \operatorname{tr}$ then $\alpha(1 \otimes p) < p$. But then $\alpha(1 \otimes p)$ is of the form $q \otimes p$ for $q \in M$, $\operatorname{tr}(q) = \lambda$. But then

$M \cong (1 \otimes p)M \otimes \mathscr{B}(\mathscr{H})(1 \otimes p) \cong (q \otimes p)M \otimes \mathscr{B}(\mathscr{H})(q \otimes p) \cong qMq$. Conversely, given an isomorphism $\theta : M \to qMq$ one may construct an automorphism of $M \otimes \mathscr{B}(\mathscr{H})$ which scales tr by $\text{tr}(q)$.

1.10. Type III factors. Although we will not spend much time on them in the lectures, it seems that this rapid survey would be absurd and misleading if we did not discuss, in the same freewheeling spirit we have established, the structure of type III factors.

Murray and von Neumann obviously considered them pathological and many problems were solved for a long time in all cases except type III. The technical problem one runs into is that if one considers the G.N.S. construction for a faithful state $\varphi : M \to \mathbb{C}$ for a type III factor M (φ weakly continuous), the mapping $* : M \to M$, which would be an isometry if φ were a trace, does not extend to an operator on the Hilbert space completion $\mathscr{H}_\varphi$ of M. It was Tomita who first used the unbounded operator S defined by $*$. One needs to extend its domain so that it is a closed operator and then one may consider the *polar decomposition* $S = J\Delta^{1/2}$ where Δ is a positive operator and J is a conjugate linear isometry. What Tomita saw and what was ultimately proved (see [**Ta1**]) was that J may be used in place of $*$. In particular, $JMJ = M'$. But it is also true that $\Delta^{it} M \Delta^{-it} = M$ (for $t \in \mathbb{R}$); so one gets a one-parameter automorphism group σ_t^φ, the modular group, straight from the state φ! Connes showed in [**Co1**] that σ_t^φ only depends on φ up to inner automorphisms so that the group $T(M) = \{t | \sigma_t^\varphi \text{ is inner}\}$ is an invariant of M itself. There are still many questions of interest about $T(M)$, but another invariant $S(M)$, defined by Connes to be the intersection of the spectra of the Δ's obtained by the above process, (minus zero) letting φ vary, is necessarily a closed multiplicative subgroup of $\mathbb{R}^+$ and hence one may classify type III factors into III_λ, $\lambda \in [0, 1]$, by:

III_0: $S(M) = \{1\}$;

III_λ, $0 < \lambda < 1$: $S(M) = \{\lambda^n | n \in \mathbb{R}\}$;

III_1: $S(M) = \mathbb{R}^+$.

To obtain an example of all this, one may look at $\bigotimes_{i=1}^\infty M_2(\mathbb{C})$ and consider on it the state φ_λ, for $0 < \lambda < 1$, given by

$$\varphi_\lambda(x_1 \otimes x_2 \otimes \cdots \otimes x_n \otimes 1 \otimes 1 \otimes \cdots) = \prod_{j=1}^n \text{trace}\left(\frac{1}{1+\lambda}\begin{pmatrix} 1 & 0 \\ 0 & \lambda \end{pmatrix} x_j\right).$$

The von Neumann algebras coming from the G.N.S. construction are the Powers factors R_λ. They were shown to be mutually nonisomorphic type III factors by Powers in [**Pow**]. The operators Δ and J can be handled by finite-dimensional calculations and one may show that the factors are of type III_λ. The modular group is just conjugation by $\bigotimes_{j=1}^\infty \exp(i\begin{pmatrix} 1 & 0 \\ 0 & \lambda \end{pmatrix})$.

An alternative construction of III_λ factors is to take an automorphism α of a II_∞ factor M, scaling the trace by λ, and forming the crossed product

$M \rtimes \mathbb{Z}$, $\mathbb{Z}$ acting via α. Connes showed that all III$_\lambda$ factors, $0 < \lambda < 1$, arise in this way and all III$_0$ factors also, if one replaces M by a II$_\infty$ nonfactor with α acting ergodically on its center.

There is no such general discrete decomposition for III$_1$ factors but Takesaki showed in [**Ta2**] that any III$_1$ factor is of the form $M \rtimes \mathbb{R}$ where M is a II$_\infty$ factor and $\mathbb{R}$ acts so as to scale the trace nontrivially.

Thus, in some sense, type III factors are reduced to type II factors and their automorphism groups.

1.11. Hyperfiniteness, R. A famous question in algebra is this: is a discrete group Γ determined up to isomorphism by the isomorphism class of the integral group ring $\mathbb{Z}\Gamma$? If we change from $\mathbb{Z}$ to $\mathbb{C}$ the answer is clearly no (witness $\mathbb{Z}/4\mathbb{Z}$ and $\mathbb{Z}/2\mathbb{Z} \oplus \mathbb{Z}/2\mathbb{Z}$) and we might be led to ask the question: to what extent does the von Neumann algebra completion vN(Γ) (see §1.5) remember Γ? There seem to be two answers to this question depending on what kind of group Γ is: not very much, and completely. The second possibility has been raised by Connes (though not yet proved), working by analogy with the Mostow rigidity result [**Mos**], for discrete cofinite subgroups of semisimple Lie groups of real rank ≥ 2. We shall concentrate on the other extreme. We will see that as soon as a group Γ is a union of finite subgroups (and is i.c.c.) then the II$_1$ factor vN(Γ) is independent of Γ.

We shall say that a von Neumann algebra M is *hyperfinite* (not wonderful terminology but it seems to have some primitive appeal) if there is an increasing sequence A_n of finite-dimensional von Neumann subalgebras of M whose union is weakly dense in M.

It is a fundamental theorem of Murray and von Neumann that there is, up to abstract algebraic isomorphism, a unique hyperfinite II$_1$ factor which we shall denote by the letter R. This is proved by a cutting and rebuilding argument which is nowadays considered standard technical machinery.

Let us reconsider the examples of §1.5 in the light of the above theorem. If Γ is the group S_∞ of finite permutations of $\mathbb{N}$, vN(Γ) is obviously hyperfinite, so $\cong R$. It follows from deep results of Connes (see §1.12) that vN(Γ) is hyperfinite as soon as Γ is amenable, i.e., there is a left-invariant mean on $l^\infty(\Gamma)$. It is not obvious but true that the II$_1$ factor $L^\infty(S^1) \rtimes \mathbb{Z}$ is also hyperfinite. In fact, the crossed product of $L^\infty(X, \mu)$ by $\mathbb{Z}$ is always hyperfinite. Much more generally, by the results of [**Co6**], $M \rtimes \Gamma$ is hyperfinite as soon as M is and Γ is amenable. The infinite tensor product algebra of §1.5 d) is obviously hyperfinite.

Are there nonhyperfinite II$_1$ factors? Let us point out a special feature of R using, say, the vN(S_∞) model. Let S_i be the transposition $(i\ i+1)$, permutation of $\mathbb{N}$. If we use the 2-norm $\|x\|_2 = \sqrt{\mathrm{tr}(x^*x)}$, it is clear that for any $y \in \mathbb{C}S_\infty$, $[S_i, y] = 0$ for large i, so that for any $y \in \mathrm{vN}(S_\infty)$, $\lim_{i \to \infty} \|[S_i, y]\|_2 = 0$. On the other hand, $\mathrm{tr}(S_i) = 0$; so S_i

stays well away from the center of $vN(S_\infty)$. Such a (norm bounded) sequence is called a central sequence (exercise: prove that all central sequences are trivial in finite dimensions). Thus R has nontrivial central sequences. On the other hand, if Γ is the free group F_2 with generators a and b, it is not hard to show that there is a constant K such that, for $y \in vN(F_2)$, $\|y - \text{tr}(y)1\|_2 < K \max\{\|[a, y]\|_2, \|[b, y]\|_2\}$. Thus any central sequence in $vN(F_2)$ is trivial. Hence $vN(F_2) \not\cong R$.

So how many II_1 factors are there? More than you care to think about. With one glorious exception due to Connes (using Kazdan's property T; see [**Co2**]), all constructions of many many II_1 factors ultimately rely on a clever manipulation of central sequences. If there are none, and property T is not around, we remain totally in the dark. It is shameful but true that we do not know if $vN(F_2) \cong vN(F_3)$! Nor do we know if the fundamental group of $vN(F_2)$ contains a single element different from 1.*

In fact all hyperfinite factors are known. The type I case is trivial, the II_1 case is given by the uniqueness result of Murray and von Neumann. Connes showed that there is only one hyperfinite factor in the cases II_∞, III_λ, $0 < \lambda < 1$, and that hyperfinite III_0 factors are classified by ergodic transformations, using work of Krieger ([**Kr**]). Haagerup proved uniqueness of the hyperfinite III_1 factor, in [**Ha**].

---

*But for F_∞ it appears that the fundamental group is $\mathbb{R}$ according to recent results of Voiculescu and Radulescu.

Lecture 2. Group Actions and Subfactors

2.1. The coupling constant. We begin by giving Murray and von Neumann's pretty but unenlightening definition of the coupling constant for a II_1 factor acting on a Hilbert space. By now we are supposed to be converted to the abstract view of a II_1 factor. The factor itself is the primary object and it may be considered as acting on different Hilbert spaces $\mathcal{H}$. We will only consider "normal" actions, for which the image of the algebra as a subalgebra of $\mathcal{B}(\mathcal{H})$ is weakly closed.

So let M be a II_1 factor on $\mathcal{H}$ with commutant M'. If M' is not of type II_1 we say the coupling constant $\dim_M(\mathcal{H})$ is infinity. Otherwise let tr_M and $\mathrm{tr}_{M'}$ be the normalized traces on M and M' respectively. Choose any nonzero vector $\xi \in \mathcal{H}$ and form the closed subspaces $\overline{M\xi}$ and $\overline{M'\xi}$, with orthogonal projections p and q, respectively. Then clearly $p \in M'$ and $q \in M$, and Murray and von Neumann showed that $\mathrm{tr}_M(q)/\mathrm{tr}_{M'}(p)$ is independent of ξ. We call this number $\dim_M(\mathcal{H})$, the coupling constant.

The first observation is that, if $\mathcal{H} = L^2(M, \mathrm{tr})$ then $\dim_M(\mathcal{H}) = 1$. In fact the converse is also true; indeed Murray and von Neumann showed that $\dim_M(\mathcal{H})$ characterizes $\mathcal{H}$ up to equivalence as an M-module.

Thus we can give a much more transparent definition of $\dim_M(\mathcal{H})$. Let K be a sufficiently large-dimensional Hilbert space and suppose M on $\mathcal{H}$ is given. Then choose an M-linear isometry $u: \mathcal{H} \to L^2(M) \otimes K$ (must exist, not necessarily surjective). We know uu^* is a projection in the commutant of M for its diagonal action on $L^2(M) \otimes K (= L^2(M) \oplus L^2(M) \oplus \cdots)$ and thus is a matrix of elements of M^{opp} (= commutant of M on $L^2(M)$). The sum of the traces of the diagonal entries of uu^* is the coupling constant $\dim_M(\mathcal{H})$. Note that if this is finite then K need only be finite-dimensional. Note also that if we had followed this procedure for $M = \mathbb{C}$ we would have obtained the usual integer $\dim_\mathbb{C}(\mathcal{H})$.

This approach shows immediately how to obtain Hilbert spaces of arbitrary real dimension ≥ 0. Just reduce $L^2(M) \otimes \mathcal{H}$ by an appropriate projection in the commutant of M. On the other hand, there is one significant fact that is trivial from the Murray-von Neumann point of view but not quite so clear in this picture and, that is, that if M and M' on $\mathscr{H}$ are both II_1 factors then $\dim_M(\mathscr{H})\dim_{M'}(\mathscr{H}) = 1$.

We end this section by pointing out that coupling constants do arise in concrete situations. For instance, if G is a locally compact group and $\mathscr{H}$ is the Hilbert space of a discrete series representation of G, then if Γ is an i.c.c. discrete subgroup of G, the von Neumann algebra M generated by Γ on $\mathscr{H}$ is a II_1 factor and $\dim_M(\mathscr{H}) =$ (covolume of Γ) × (formal dimension of $\mathscr{H}$). For terminology see [**Rob**]. In particular, all the conditions are satisfied in the case $G = PSL_2(\mathbb{R})$, $\Gamma = PSL_2(\mathbb{Z})$, and G acts in the usual way on L^2-holomorphic functions on the upper half-plane with respect to the measure $dx\,dy/y^{2-n}$. It is intriguing that the cusp forms of number theory (see [**Ser**]) are naturally interpreted in this setting as M-linear maps between the various Hilbert spaces. See [**GHJ**].

2.2. Galois theory for finite group actions. We begin by discussing some rather easy abstract nonsense results for actions of finite groups on II_1 factors. The main trick used will be the simple observation that, for a factor M and an automorphism $\alpha \in \text{Aut}(M)$, the "freeness" condition for α (§1.5, b(i)) is equivalent to α being outer. To see this, if α is outer let x be such that $x\alpha(y) = yx$ for all y in M. Then also $x^*y = \alpha(y)x^*$ for all y in M and x^*x, xx^* belong to the center of M. They are thus positive scalars; so x may be rescaled to be a unitary and $\alpha(y) = x^*yx$, i.e., α is inner, a contradiction unless $x^*x = 0$ or $xx^* = 0$, i.e., $x = 0$. The following three results are easy consequences.

A) If Γ is a discrete group acting on a factor N by outer automorphisms (for nonidentity elements) on N then $N \rtimes G$ is a factor and $N' \cap N \rtimes G = \mathbb{C}\,\text{id}$.

B) If N acts on a Hilbert space $\mathscr{H}$ and Γ is as in A) and *finite*, then if $\gamma \mapsto v_\gamma$ is a unitary representation of Γ on $\mathscr{H}$ with $v_\gamma x v_\gamma^{-1} = \gamma(x)$ for $x \in N$, then the map $x \mapsto x$, $u_\gamma \to v_\gamma$ defines an isomorphism from $N \rtimes \Gamma$ to $\{N, \{v_\gamma\}\}''$ (see [**Au**]).

The following result is a simple result if one accepts that II_1 factors have no nontrivial strongly closed ideals.

C) If N and Γ are as in B) and M is a sub*-algebra of $N \rtimes \Gamma$ containing N then there is a subgroup Γ_0 of Γ with $M = N \rtimes \Gamma_0$. (Prove it by reducing the support of $\sum a_g u_g$ to a single element.)

Combining A), B), and C), and using the action of a II_1 factor M on $L^2(M, \text{tr})$ we see that we have the following result.

THEOREM. *Let M be a II_1 factor and G a finite group of automorphisms of M, outer except for the identity. Then there is a Galois correspondence* between subgroups of G and subalgebras of M containing the fixed point algebra M^G, given by $H \leftrightarrow M^H$. Moreover, $(M^G)' \cap M = \mathbb{C}\,\mathrm{id}$.

PROOF. To prove this simply let M act on $L^2(M, \mathrm{tr})$. Then each $g \in G$ extends to a unitary v_g on the Hilbert space with $v_g x v_g^* = g(x)$ for $x \in M$. The v_g's thus also normalize $M' = JMJ$ and are outer on it; so $(M^G)' = \{M' \cup \{v_g\}\}''$ is a II_1 factor $\cong M' \rtimes G$. Obviously the commutant sets up a Galois correspondence between subalgebras intermediate between M^G and M and subalgebras between M' and $(M^G)'$; so we are done.

These results should be attributed to the Japanese school ([NT], [Suz]).

Although rather elementary, the above results, and the method used to prove them, were crucial motivation for the consideration of index for subfactors to follow.

2.3. Connes' results on automorphisms of R.
In two remarkable papers ([Co3] and [Co4]), Connes completely classified automorphisms of the hyperfinite II_1 and II_∞ factors up to outer conjugacy.

DEFINITION. Two automorphisms α and β of a von Neumann algebra M are called *conjugate* if there is an automorphism θ of M with $\theta \alpha \theta^{-1} = \beta$. They are called *outer conjugate* if there is a unitary $u \in M$ such that $(\mathrm{Ad}\,u)\alpha$ is conjugate to β where $\mathrm{Ad}\,u$ is the inner automorphism $\mathrm{Ad}\,u(x) = uxu^*$.

Connes' classification is as follows.

1) *Outer conjugacy*. If $\alpha \in \mathrm{Aut}\,M$ is an automorphism of a factor M, let $p = p(\alpha)$ be the smallest integer ≥ 1 such that α^p is an inner automorphism. Then if $\alpha^p = \mathrm{Ad}\,u$, $\alpha(u) = \gamma u$ for a pth root of unity γ. The pair (p, γ) is a complete invariant of outer conjugacy of automorphisms of the hyperfinite II_1 factor R. Moreover, for each pair (p, γ), $\gamma \in \mathbb{C}$, $\gamma^p = 1$, there is an automorphism s_p^γ having (p, γ) as its invariants.

If M is a II_∞ factor and $\alpha \in \mathrm{Aut}\,M$, then there is a real number $\lambda > 0$ such that $\mathrm{tr} \circ \alpha = \lambda \,\mathrm{tr}$ for any trace (not everywhere defined) on M. The number λ is an outer conjugacy invariant of α called $\mathrm{mod}(\alpha)$. For each $\lambda > 0$ there is an automorphism α of the hyperfinite II_∞ factor $R_{0,1}$ with $\mathrm{mod}(\alpha) = \lambda$, and such an α is unique up to outer conjugacy if $\lambda \neq 1$. If $\mathrm{mod}(\alpha) = 1$ the invariants (p, γ) defined above are a complete set of outer conjugacy invariants for automorphisms of $R_{0,1}$.

2) *Conjugacy*. If two automorphisms α and β are outer conjugate and periodic (to be conjugate they have to have the same period), say $\alpha^n = \beta^n = 1$, and if $\alpha^p = \mathrm{Ad}\,u$ as above, a new invariant enters via the traces of the spectral projections of u. It is a probability measure defined on the n/pth roots of unity, up to rotation, and Connes showed that this invariant, together with (p, γ) as above, is a complete conjugacy invariant.

If α and β are not periodic, the conjugacy problem is still very complicated and many invariants (e.g. entropy, see [**CS**]) are available.

If M is $R_{0,1}$ and $\mod(\alpha) \neq 1$, outer conjugacy actually implies conjugacy ([**Co4**]).

2.4. Extensions of Connes' automorphism results. There are two ways in which one could extend the results of §2.3 for II_1 factors. One could first consider automorphisms of nonhyperfinite factors M. Connes showed that the results can be extended somewhat if α is in the closure of the inner automorphism group and if α acts nontrivially on central sequences. Such reflections led Connes to his beautiful invariant $\chi(M)$ for II_1 factors (see [**Co5**]). If α is not in the closure of the inners, the situation is hopeless as shown by Phillips in [**Ph**].

Another approach to generalization is to stay with R and $R_{0,1}$ and replace $\mathbb{Z}$ and $\mathbb{Z}/n\mathbb{Z}$ by more general groups. The first thing to be done is to generalize the invariants described above. If G is a group and $\alpha: G \to \operatorname{Aut} M$ is an action of G on a factor, define $N(\alpha)$ to be $\alpha^{-1}(\operatorname{Int} M)$, a normal subgroup, and for each h in $N(\alpha)$ choose a unitary u_h with $\alpha_h(x) = u_h x u_h^*$. Then it is easy to show there are functions $\lambda: G \times N(\alpha) \to \mathbb{C}$ and $\mu: N(\alpha) \times N(\alpha) \to \mathbb{C}$ with

$$\alpha_g(u_h) = \lambda(g, h) u_{ghg^{-1}},$$
$$u_h u_k = \mu(h, k) u_{hk}.$$

The pair (λ, μ) satisfies certain relations and is only defined modulo changing the choice of the u_h's. Thus it naturally lives in some cohomology group $H^2(G; N(\alpha))$, which turns out to be the relative cohomology so that the sequence $(Q = G/M)$

$$0 \to H^1(Q) \to H^1(G) \to H^1(N(\alpha)) \to H^2(Q) \to H^2(G)$$
$$\to H^2(G; N(\alpha)) \to H^3(Q) \to H^3(G)$$

is exact (circle coefficients throughout). The pair (λ, μ) in $H^2(G; N(\alpha))$ is called the *characteristic invariant* $\Lambda(\alpha)$ (see [**J1**]), and $(N(\alpha), \Lambda(\alpha))$ generalizes the pair (p, γ) defined above. In [**Oc1**], Ocneanu proves that if G is discrete and amenable, $(N(\alpha), \Lambda(\alpha))$ is a complete invariant for "cocycle conjugacy" of actions on R, which is a slightly finer notion than outer conjugacy. If the group is finite, an inner invariant coming from traces can be defined and in [**J2**] it is shown that all these invariants give a complete conjugacy invariant. In particular, if the action of G is outer, there is only one action of G, up to conjugacy, on the hyperfinite type II_1 (or II_∞) factor. By the results of §2.3, one has a whole family of subfactors, namely fixed point algebras for actions of finite groups, which are completely parametrized up to conjugacy by finite groups.

If the group is not discrete much less information is available. For compact abelian groups the problem can be solved by duality (see [**JT**]). For compact nonabelian groups the interesting question arises as to whether simple groups can act ergodically (fixed points = scalars) on R. (If it is possible on an infinite-dimensional factor it must be R—see [**HLS**].) The answer is no for SU(2) as shown by Wassermann in [**Wa**].

2.5. Index for subfactors. Let us return now to the Galois theory for finite group actions from the point of view of one interested in subfactors. It is easy to show that the normalizer $N(M)$ of M in $M \rtimes G$ ($N(M) = \{u \in M \rtimes G \mid u$ unitary, $uMu^* = M\}$) is the semidirect product $U(M) \rtimes G$ (provided the action of G is outer) so that the subfactor $M \subset M \rtimes G$ remembers the group. Using the theory of §2.2 one can also see that there is a nice little theory for subfactors related to the crossed product analogous to subfields in Galois theory. What is missing is the degree of an extension. But the coupling constant furnishes us with that. It is easy to show that if M acts on $L^2(M)$ and we realize $M \rtimes G$ on $\mathcal{H} = L^2(M) \otimes l^2(G)$ as in §1.5, then the coupling constants of M and $M \rtimes G$ on $\mathcal{H}$ are $|G|$ and 1 respectively. Thus if we defined the degree of the extension by the ratio of the coupling constants we would get exactly the right properties. For the pair $M \rtimes H \subseteq M \rtimes G$ the degree will obviously be $[G:H]$.

Thus we make the following definition.

DEFINITION. If $N \subseteq M$ are II$_1$ factors with the same identity, $[M:N] \stackrel{\text{def}}{=} \dim_N(L^2(M, \text{tr}))$, or $\dim_N(\mathcal{H})/\dim_M(\mathcal{H})$ whenever $\mathcal{H}$ is a Hilbert space on which M acts with $\dim_M(\mathcal{H}) < \infty$. This number is called the *index* of N in M. (The definition works in finite dimensions also.)

This puts the cap on the Galois theory and one begins to suspect that all subfactors may be obtainable from group actions, at least in finite index. The suspicion is strengthened by the following result due to M. Goldman [**Go**].

THEOREM. *Let $N \subseteq M$ be II$_1$ factors with $[M:N] = 2$. Then there is an outer $\mathbb{Z}/2\mathbb{Z}$ action on N for which the pair $N \subseteq M$ is isomorphic to the pair $N \subseteq N \rtimes \mathbb{Z}/2\mathbb{Z}$.*

Before going on we gather together some simple results about the index, which are all trivial manipulations of the coupling constant.

(i) If $N \subseteq M \subseteq P$ then $[P:N] = [P:M][M:N]$.
(ii) If $N \subseteq M$ act on $\mathcal{H}$ and N' is type II$_1$ then $[N':M'] = [M:N]$.
(iii) If p is a projection in $N' \cap M$ then ($[M:N] < \infty$) $[pMp:pNp] = [M:N]\text{tr}_M(p)\text{tr}_{N'}(p)$.
(iv) If $N_1 \subseteq M_1$, $N_2 \subseteq M_2$ then $[M_1 \otimes M_2 : N_1 \otimes N_2] = [M_1:N_1][M_2:N_2]$.

The relative commutant $N' \cap M$ will play an increasing role and we add to our list of properties.

(v) If $[M:N] < \infty$ then $\dim_{\mathbb{C}}(N' \cap M) < \infty$.
(vi) If $\dim(N' \cap M) > 1$, $[M:N] \geq 4$.

The last deserves a proof: since $N' \cap M$ is a von Neumann algebra, it contains a projection $p \neq 0$ or 1. Since $[pMp : pNp] \geq 1$ we have

$$[M:N]\mathrm{tr}_M(p) \geq 1/\mathrm{tr}_{N'}(p), \qquad [M:N]\mathrm{tr}_M(1-p) \geq 1/\mathrm{tr}_{N'}(1-p).$$

Adding we get $[M:N] \geq 1/d + 1/(1-d)$ for some d, $0 < d < 1$.

Note that, besides the very algebraic definition of $[M:N]$, another purely analytic one is given by Pimsner and Popa in [**PP1**]. This has proved extremely useful, initially, in their remarkable insight that index and Connes-Stormer entropy are related. Here is a version of their definition:

$$[M:N]^{-1} = \inf\{\lambda \mid \exists \text{ a projection } p \text{ in } M, \ E_N(p) \leq \lambda \, \mathrm{id}\}.$$

2.6. The basic construction. We now come to the main ingredient of this work, the thing that makes everything go. It was also discovered and exploited in different ways by Skau [**Sk**] and Christensen [**Ch**]. For myself its importance was suggested by the answer to the following question: in $M \rtimes G$ ($|G| < \infty$) the projection $e = (1/|G|) \sum_{g \in G} u_g$ plays a special and important role. What is the von Neumann algebra generated by M and e? The answer, as we shall see, is "all of $M \rtimes G$". At the time this seemed surprising as it is not at all clear how to write, say, u_g as an algebraic combination of e and elements of M.

If $M \rtimes G$ is realized on $L^2(M)$ in canonical fashion as in §1.2, then e is just the extension to $L^2(M)$ of the so-called "conditional expectation" $E: M \to M^G$. This conditional expectation always exists ([**U**]) for a sub von Neumann algebra N of a II$_1$ factor M and has the following properties which also define it.

(i) $E(axb) = aE(x)b$ if $a, b \in N$, $x \in M$.
(ii) $\mathrm{tr}(aE(x)) = \mathrm{tr}(ax)$ for $a \in N$, $x \in M$.
(iii) $E(x^*) = E(x)^*$ and $E(x^*x) \geq E(x)^*E(x)$.

All this discussion leads to the following

DEFINITION. Let $N \subseteq M$ be II$_1$ factors. The *basic construction* is to let M act on $L^2(M)$ and construct the von Neumann algebra $\langle M, e \rangle$ on $L^2(M)$ generated by M and the extension to $L^2(M)$ of the conditional expectation E of M onto N.

Here are some elementary facts about the basic construction.
(a) If $x \in M$, $exe = E(x)e$.
(b) $\{e\}' \cap M = N$.
(c) If J is the involution of $L^2(M)$ as in §1.8, then $JeJ = e$ and $N' = J\langle M, e\rangle J$, or $\langle M, e\rangle = JN'J$.
(d) Linear combinations $\sum_i x_i e y_i$, $x_i, y_i \in M$, are a dense *-subalgebra of $\langle M, e\rangle$.

(e) If $[M:N] < \infty$ then $\langle M, e_N \rangle$ is a II_1 factor and
$$[\langle M, e_N \rangle : M] = [M:N].$$

(f) $\text{tr}(e)[M:N] = 1$.

(g) If $x \in M$, $\text{tr}(xe) = \text{tr}(x)\text{tr}(e) = \text{tr}(x)[M:N]^{-1}$.

These are all straightforward though the bicommutant theorem is used somewhere, probably in (c), and (d) uses (ii) of §2.5, which in turn uses $\dim_M(\mathcal{H}) \dim_{M'}(\mathcal{H}) = 1$ (see §2.1). Let us prove (g): $\text{tr}(xe) = \text{tr}(exe) = \text{tr}(E(x)e)$, but $z \mapsto \text{tr}(ze)$ is a trace on N, so by uniqueness of the trace, $\text{tr}(E(x)e) = \text{tr}(E(x))\text{tr}(e) = \text{tr}(x)\text{tr}(e)$.

The usefulness of the basic construction is immediate. Here is the first step on the way to proving the index is an integer.

LEMMA. *If $N \subseteq M$ are II_1 factors then $[M:N]$ cannot be strictly between 1 and 2.*

PROOF. Let us calculate the index of $(1-e)N$ in $(1-e)\langle M, e\rangle(1-e)$. By (ii) of §2.5 and (f) above, it is $[M:N]^2(1-[M:N]^{-1})\text{tr}_{N'}(1-e)$; but by (a) above $e\langle M, e\rangle e = Ne$; so again by (ii) of §2.5, $1 = [M:N]^2 \text{tr}_M(e)\text{tr}_{N'}(e)$; so $\text{tr}_{N'}(1-e) = 1 - \text{tr}_M(e)$. Combining we get

$$[(1-e)\langle M, e\rangle(1-e) : N(1-e)] = ([M:N]-1)^2$$

which is less than 1 if $[M:N]$ is between 1 and 2.

We shall exploit the basic construction enormously; so we immediately make it more concrete by describing how it works in finite dimensions.

2.7. The basic construction in finite dimensions. There is no reason why we should not perform the basic construction for any finite von Neumann algebra M and sub-von Neumann algebra N provided M has a faithful (weakly continuous) trace ($\text{tr}(x^*x) = 0 \Rightarrow x = 0$). We will have to be more careful about which trace we are talking about as traces are not unique in non-factors. Thus properties (e), (f), and (g) of §2.6 do not make sense.

In particular, M and N might be finite-dimensional algebras. But the structure of such von Neumann algebras is trivial, so it is of interest to analyze the situation completely. We come to the elementary but very useful notion of a Bratteli diagram and we strongly recommend the reader to do a few exercises to become familiar with it. The idea is first to represent a finite-dimensional von Neumann algebra by a row of integers which are the ranks of the matrix algebra summands. Thus 5 4 2 is shorthand for $M_5(\mathbb{C}) \oplus M_4(\mathbb{C}) \oplus M_2(\mathbb{C})$. Then a von Neumann subalgebra N of M is represented above (or below) the row of N by its own numbers. The numbers on the two rows are connected by lines whose significance is the only nonobvious thing in the discussion. We give two concrete examples which should serve to make the meaning of the lines (multiplicities) clear.

EXAMPLE. $\overset{1}{\underset{1}{\diagup}}\overset{2}{\underset{1}{\diagdown}}$ is the Bratteli diagram for the pair $N \subseteq M$ where $M = M_2(\mathbb{C}) \oplus \mathbb{C}$ and N is the subalgebra $\{(\begin{smallmatrix} a & 0 \\ 0 & b \end{smallmatrix}) \oplus b | a, b \in \mathbb{C}\}$.

EXAMPLE. $\overset{5}{\underset{2\,\,1}{\diagup\diagdown}}$ is the Bratteli diagram of the pair $N \subseteq M$ where $M = M_5(\mathbb{C})$ and N is the subalgebra

$$\left\{ \begin{pmatrix} A & 0 & 0 \\ 0 & A & 0 \\ 0 & 0 & x \end{pmatrix} \middle| A \in M_2(\mathbb{C}),\, x \in \mathbb{C} \right\}.$$

We leave it to the reader to give a precise definition of the multiplicities.

Note that the "addition rule" $5 = 2 \times 2 + 1$, etc., in the above examples simply expresses the fact that N and M have the same identity.

Now what happens if we do the basic construction to a pair $N \subseteq M$ as above? Certainly $\langle M, e_N \rangle$ is a finite-dimensional von Neumann algebra. Moreover, by (c) of §2.6, its center can be identified with that of N; so the Bratteli diagram for $M \subseteq \langle M, e \rangle$ must consist of the same set of points for M connected somehow to the same set of points for N. The only reasonable way to do this is to have the same connection pattern. It is an amusing exercise to prove this. Thus for the two examples above we will have

$$\begin{array}{c} \overset{3}{}\overset{2}{} \\ 1\,\,2 \\ 1\,\,1 \end{array} \begin{array}{c} \leftarrow \langle M, e \rangle \\ \leftarrow M \\ \leftarrow N \end{array}$$

and

$$\begin{array}{c} 10\,\,\,\,\,5 \\ 5 \\ 2\,\,\,\,\,1 \end{array} \begin{array}{c} \leftarrow \langle M, e \rangle \\ \leftarrow M \\ \leftarrow N. \end{array}$$

Note that the answer is independent of the faithful trace used in N.

2.8. Two basic constructions, proof of Goldman's theorem. We once again suppose that M and N are factors, $[M:N] < \infty$.

In §2.6 we saw that the basic construction immediately implies $[M:N]$ cannot be between 1 and 2. With this success it was tempting to try it again, i.e., repeat it for the pair $M \subseteq \langle M, e \rangle$. We would then put $e = e_1$ and the "e" coming from the conditional expectation onto M we would call e_2. Then we have the situation $N \subseteq M \subseteq \langle M, e_1 \rangle \subseteq \langle M, e_1, e_2 \rangle$, though note that we change Hilbert spaces going from $\langle M, e_1 \rangle$ to $\langle M, e_1, e_2 \rangle$. The following relations are important.

Put $\tau = [M:N]^{-1} = \text{tr}(e_1) = \text{tr}(e_2)$. Then

(i) $e_2 e_1 e_2 = \tau e_2$;
(ii) $e_1 e_2 e_1 = \tau e_1$.

Relation (i) follows from (a) of §2.6 but (ii) is slightly more subtle and requires (d) of §2.6.

These relations make the determination of the algebra generated by 1, e_1, and e_2 easy. It is at most five-dimensional, for the set $\{1, e_1, e_2, e_1e_2, e_2e_1\}$ is multiplicatively closed modulo scalars. It is easy to show that $(e_1 - e_2)^2/(1-\tau)$ is a projection. With this we can prove Goldman's theorem.

PROOF OF GOLDMAN'S THEOREM. If $\tau = \frac{1}{2}$, i.e., $[M:N] = 2$, then $\operatorname{tr}(e_1-e_2)^2/(1-\tau) = 2(\operatorname{tr}(e_1) + \operatorname{tr}(e_2) - 2\operatorname{tr}(e_1 e_2)) = 2(1-\frac{1}{2}) = 1$, so that $e_1 + e_2 - e_1 e_2 - e_2 e_1 = \frac{1}{2}$, or $(2e_2-1)e_1(2e_2-1) = (1-e_1)$. Thus if $u = (2e_2-1) \in \langle M, e_1, e_2 \rangle$ we have $u^2 = 1$ and $u(\sum x_i e_1 y_i)u^{-1} = \sum x_i(1-e_1)y_i$ for $x_i, y_i \in M$ (by (b) of §2.6). Thus by (d) of §2.6, $u\langle M, e_1\rangle u^{-1} = \langle M, e_1\rangle$, which is more than enough to prove that $\langle M, e_1, e_2\rangle = \langle M, e_1\rangle \rtimes \mathbb{Z}/2\mathbb{Z}$. To conclude that $M = N \rtimes \mathbb{Z}/2\mathbb{Z}$ one can use duality ([**Ta2**]) or universality of the basic construction (any subfactor of finite index is of the form $M \subseteq \langle M, e_1\rangle$) to find the appropriate e_1 and e_2 inside M.

Note that an important principle has been used here: the actual dimension of the algebra generated by 1, e_1, and e_2 is 4 in index 2. It is determined by positive definiteness of the trace. The element $1 - (e_1-e_2)^2/(1-\tau)$ is a "ghost" when $\tau = 1/2$.

Lecture 3. Values of the Index, Virasoro Algebra

3.1. Values of the index. We have been leading the reader to believe that $[M:N]$ is necessarily an integer. However, given the proof of Lemma 2.7 and (iii) of 2.6, one could not miss the first construction of §3.3, which will show how to construct subfactors of R with arbitrary real index ≥ 4. The next theorem was something of a surprise.

THEOREM ([**J3**]). *Let $N \subseteq M$ be II_1 factors with the same identity. Then*

(a) *if $[M:N] < 4$ there exists an integer $n \geq 3$ with $[M:N] = 4\cos^2 \pi/n$.*

(b) *For each integer $n \geq 3$ there is a subfactor R_0 of R with $[R:R_0] = 4\cos^2 \pi/n$ (R is the hyperfinite type II_1 factor).*

(c) *For any real $r \geq 4$, there is a subfactor R_0 of R with $[R:R_0] = r$.*

Thus the set of all possible index values is as below:

```
    |       |     |   | | | |  ├─────────────
    1       2    φ²   3       4
```

where $\varphi^2 = \frac{3+\sqrt{5}}{2} \approx 2.6180339$.

We shall prove this theorem by the original method with an improvement due to Wenzl ([**We1**]).

If one moves away from the hyperfinite II_1 situation things change. It is a remarkable result of Pimsner and Popa ([**PP2**]) that if M has property T of Connes ([**CJ**]) then the set of index values is countable. All known constructions of the $4\cos^2 \pi/n$ series (except $n = 3, 4$, and 6) rely heavily on hyperfiniteness; so it remains an open question to find examples of these values for II_1 factors M without MacDuff's property $M \cong M \otimes R$ (see [**McD**]). It might be possible to get contrived examples but one would hope for a genuinely new construction.*

There is an index theory for type III factors developed by Kosaki [**Kos**]. The index then requires (and depends upon) a normal conditional expectation

---

*Popa has since solved this problem.

onto the subfactor. Theorem 3.1 holds in this context, but Loi has shown in [**Lo**] that the index question seems to break down to a II_1 question.*

The spectrum of index values is known to occur in other places. For many occurrences see [**GHJ**]. We mention here only the following, associated with Hecke [**He**]: consider the subgroup of $SL_2(\mathbb{R})$ generated by $\begin{pmatrix} 1 & \lambda \\ 0 & 1 \end{pmatrix}$ and $\begin{pmatrix} 0 & 1 \\ -1 & 0 \end{pmatrix}$. For what values of $\lambda > 0$ is it discrete? Answer: $\lambda = 2\cos\pi/n$, $n = 3, 4, \ldots$ or $\lambda \geq 2$. The case $\lambda = 1$ (which would correspond to a trivial subfactor) gives the group $SL_2(\mathbb{Z})$. We have been unable to find any direct connection between this result and Theorem 3.1. It is a tantalizing situation.

3.2. The Virasoro unitarity result. If a theory involves the circle we may expect the group of diffeomorphisms of the circle, $\mathrm{Diff}(S^1)$, to play a role even if only as reparametrizations. In quantum physics we expect groups to be represented by unitaries on the underlying Hilbert space, although one can only deduce the existence of a projective representation on abstract grounds. Thus one will expect the study of projective unitary representations of $\mathrm{Diff}(S^1)$ to be of some interest in physics. We consider a problem that is simplified in two ways—first we replace $\mathrm{Diff}(S^1)$ by its complexified Lie algebra; second we consider only those vector fields that are trigonometric polynomials, i.e., finite linear combinations of $e^{in\theta}d/d\theta$ where θ is the angle variable on the circle. This Lie algebra is obviously defined by the relation $[e^{im\theta}\frac{d}{d\theta}, e^{in\theta}\frac{d}{d\theta}] = i(n-m)e^{i(m+n)}\frac{d}{d\theta}$. A projective unitary representation of $\mathrm{Diff}(S^1)$ will define a central extension of this Lie algebra and it is not difficult to see that the most general such extension is the Lie algebra Vir with basis $\{L_n : n \in \mathbb{Z}\} \cup \{C\}$ and Lie bracket

$$[L_n, L_m] = (n-m)L_{m+n} + \delta_{n,-m}\frac{(n^3-n)}{12}C,$$
$$[C, L_n] = 0.$$

(The map $L_n \to (1/i)e^{in\theta}\frac{d}{d\theta}$ has kernel spanned by C and defines the universal central extension.)

The element C is called the central charge and in an irreducible representation will be a scalar multiple of the identity; so we will talk about it as a number.

In a projective unitary representation of $\mathrm{Diff}(S^1)$ the L_n's will still be (complexified) generators of one-parameter groups; so the unitarity condition imposes $L_n^* = L_{-n}$ (at least in a formal sense; we do not consider domains).

A third simplification we shall make is to suppose that our projective unitary representation is irreducible and has a lowest weight vector $|h\rangle$, i.e., a unit vector such that $L_n|h\rangle = 0$ for $n > 0$, and $L_0|h\rangle = h|h\rangle$. It is then clear that the linear space V spanned by $\{L_{i_1}L_{i_2}\cdots L_{i_k}|h\rangle : i_1, i_2, \ldots, i_k < 0\}$ is

---

* Though not in the III_0 case. See [**Kos 1**].

a dense subspace of the Hilbert space and that the parameters h and C completely determine the representation. The following result is due to Friedan, Qiu, and Schenker [**FQS**].

THEOREM. *Given an irreducible unitary representation of* Vir *with central charge C and highest weight h, then $C \geq 0$, $h \geq 0$, and if $C < 1$ there is an integer $m \geq 2$ for which $C = 1 - \frac{6}{m(m+1)}$. If C is $1 - \frac{6}{m(m+1)}$ then there is a $p \in \{1, 2, \ldots, m-1\}$ and $q \in \{1, 2, \ldots, p\}$ for which $h = \frac{[(m+1)p - nq]^2 - 1}{4m(m+1)}$. If $C > 1$ there are such representations of* Vir *for any C and h.*

We shall see how the proof proceeds in §3.7. For the moment we would just like to remark on the similarity between Theorems 3.1 and 3.2. Of course there are many areas in mathematics and physics which have both discrete and continuous spectrum; so it would be foolish to take the above as evidence of a profound connection. But we would like to develop the two theories in parallel with the intention of convincing the audience that there is indeed a remarkable analogy between them. It is hoped that this analogy will eventually prove fruitful to both theories.

3.3. The continuous series for subfactors. We give two related constructions of the continuous series of index values ≥ 4.

Construction A. The proof of (vi) of §2.5 shows how to get examples: if the subfactor N had the property $\dim(N' \cap M) = 2$, and if p is a nonzero projection in $N' \cap M$ with both $pMp = Np$ and $(1-p)M(1-p) = N(1-p)$, then we would have $[M : N] = \operatorname{tr}_M(p)^{-1} + (1 - \operatorname{tr}_M(p))^{-1}$. The way to construct such a subfactor is to use the fundamental group, which for R is $\mathbb{R}$; so there is no problem. Just choose a projection $p \in R$ of trace d and an isomorphism $\theta : pRp \to (1-p)R(1-p)$. Then $\{x + \theta(x) | x \in pRp\}$ is obviously a subfactor which equally obviously has the correct property.

Construction B. This method is much more interesting although the subfactors it produces are just the same as the previous ones. However, the idea has a big role to play in subfactors.

We start with the Powers factor R_λ of type III_λ, which is obtained as in §1.10 by applying the G.N.S. construction to $\bigotimes_{i=1}^\infty M_2(\mathbb{C})$ with respect to the product state φ_λ where $\varphi_\lambda(x) = \operatorname{trace}((\bigotimes_{i=1}^\infty d)x)$ where $d = \frac{1}{1+\lambda}\begin{pmatrix} 1 & 0 \\ 0 & \lambda \end{pmatrix}$. The fixed point algebra for the modular group is isomorphic to R. Moreover, the modular group $\sigma_t^{\varphi_\lambda}$ respects the tensor product decomposition; so we may consider the subfactor of R defined by the image of the obvious shift endomorphism of R_λ, restricted to R. We claim that the index of this subfactor is $2 + \lambda + \lambda^{-1}$. In fact, this is quite obvious, for the projection $p = \begin{pmatrix} 1 & 0 \\ 0 & 0 \end{pmatrix} \otimes 1 \otimes 1 \otimes \cdots$ is in the centralizer of φ_λ and it clearly has the reduction property used in construction A.

Thus the continuous series is rather easy to construct. But it is also clear that $N' \cap M \neq \mathbb{C}$. If one imposes on subfactors the irreducibility condition $N' \cap M = \mathbb{C}$ then it is not expected to find continuous variation for the index. We will take this point up again in §4.7.*

3.4. Iterating the basic construction: the e_i algebra. Having already profited from repeating the basic construction once it would seem foolhardy not to look for more by doing it again and again. Thus starting from a pair $N \subseteq M$ one obtains an increasing sequence of II_1 factors M_i defined by the second-order difference equation $M_{i+1} = \langle M_i, e_i \rangle$ where e_i is the extension to $L^2(M_i)$ of the conditional expectation from M_i to M_{i-1}. The initial conditions are $M_0 = N$, $M_1 = M$. One could also have obtained an equivalent structure inside the algebra M, but it would not have been canonical. The tower M_i is canonical and any structure it has is an invariant for the pair $N \subseteq M$. Note that by uniqueness of the trace on a II_1 factor there is a unique trace tr on $\bigcup_i M_i$.

We shall be especially interested in the sequence e_i of projections in the tower. The following relations are immediate (remember $\tau = [M:N]^{-1}$):

(i) $e_i^2 = e_i$, $e_i^* = e_i$,
(ii) $e_i e_{i\pm 1} e_i = \tau e_i$,
(iii) $e_i e_j = e_j e_i$ if $|i-j| \geq 2$,
(iv) $\text{tr}(x e_{n+1}) = \tau \, \text{tr}(x)$ if x is in the algebra generated by $1, e_1, e_2, \ldots, e_n$.

Property (i) comes from the conditional expectation properties.
Property (ii) comes from (i) and (ii) of §2.8.
Property (iii) comes from (b) of §2.6.
Property (iv) comes from (g) of §2.6, which in turn followed from uniqueness of the trace on a II_1 factor.

We will get a lot of mileage from these e_i's. We would like to emphasize right now that what we mean by "the e_i algebra" is the algebra in the tower M_i generated by the e_i's in the tower. We will see that the algebra depends only on $[M:N]$ but that it does vary a lot with $[M:N]$. When $[M:N] \geq 4$ one obtains the so-called Temperley-Lieb algebra which we shall study in §8.3.

DEFINITION. The e_i *algebra* $\mathscr{A}_{n,\tau}$ is the algebra generated by $\{1, e_1, e_2, \ldots, e_n\}$ inside the tower M_i defined above if $[M:N] = \tau^{-1}$.

3.5. Combinatorics of the e_i's. The first main observation about the e_i algebra is that it is finite dimensional if only finitely many e_i's are considered. This follows from (v) of §2.5 but it is instructive to prove it directly from the relations of §3.4. To get started, notice that we proved this already for $\mathscr{A}_{2,\tau}$ in §2.8 when we showed that $\{1, e_1, e_2, e_1 e_2, e_2 e_1\}$ was essentially multiplicatively closed. If we try the same naive procedure for $\mathscr{A}_{3,\tau}$ we

---
* Popa has recently solved this problem.

obtain the list $\{1, e_1, e_2, e_3, e_1e_2, e_1e_3, e_2e_1, e_2e_3, e_3e_2, e_1e_2e_3, e_3e_2e_1, e_2e_1e_3e_2, e_2e_1e_3, e_1e_3e_2\}$, a total of 14 elements. Thus we have generated multiplicatively closed sets of sizes: 2 for $\mathscr{A}_{1,\tau}$, 5 for $\mathscr{A}_{2,\tau}$, 14 for $\mathscr{A}_{3,\tau}$, and a little work gives 42 for $\mathscr{A}_{4,\tau}$. Any competent combinatorist knows that the next number in the sequence is 132 and that we are talking about the Catalan numbers $\frac{1}{k+1}\binom{2k}{k}$. These are well known to count many combinatorial objects such as triangulations of a polygon, bifurcating trees, and random walks not crossing the diagonal on $\mathbb{N} \times \mathbb{N}$. We are soon led to a general multiplicatively closed set for $\mathscr{A}_{n,\tau}$ consisting of "increasing products of decreasing strings". One example should suffice to explain this term:

$$(e_4e_3e_2)(e_5e_4)(e_7e_6e_5)(e_8).$$

The lowest index of the decreasing strings is also increasing and it is not hard to find a bijection with the random walks not crossing the diagonal, or to count the possibilities directly. We find that we have proved

$$\dim \mathscr{A}_{n,\tau} \leq \frac{1}{n+2}\binom{2n+2}{n+1}.$$

Remember we have no special reason to think that $\dim \mathscr{A}_{n,\tau} = \frac{1}{n+2}\binom{2n+2}{n+1}$; indeed we have already seen that $\dim \mathscr{A}_{2,1/2} = 4$ rather than 5. In fact relations (i)–(iv) of §3.4 determine the algebra structure completely since, as we shall see, they suffice to calculate the trace of any element of $\mathscr{A}_{\tau,n}$, and hence to decide if x is zero or not by calculating $\operatorname{tr}(x^*x)$. Thus the following result is important.

LEMMA. *Let* $\varphi : \mathscr{A}_{n,\tau} \to \mathbb{C}$ *be a linear function satisfying* $\varphi(ab) = \varphi(ba)$. *Then* φ *is completely determined by its effect on products* $e_{i_1} \cdots e_{i_k}$ *where all the e's commute with each other.*

PROOF. It suffices to show that any word on the e_i's in its standard form $(e_{k_1} \cdots e_{k_1-j_1}) \cdots (e_{k_p} \cdots e_{k_p-j_p})$ can be reduced to a product as in the statement of the lemma if we allow cyclic permutations as well as relations (i), (ii), and (iii) of §3.4. This is an amusing exercise which the reader will not find difficult.

We add the following result which will be useful in §8.6.

LEMMA. *The projection* $p_n = e_1e_3e_5 \cdots e_{2n-1}$ *is minimal in* $\mathscr{A}_{2n,\tau}$, *i.e.,* $p_n\mathscr{A}_{2n,\tau}p_n = \mathbb{C}p_n$.

PROOF. It is psychologically easier to prove that $e_2e_4 \cdots e_{2n}$ is minimal, which is easy and equivalent.

3.6. The e_i algebra is a II_1 factor. Given $N \subseteq M$ as usual we may form the tower M_i and let M_∞ be the II_1 factor obtained by applying the G.N.S. construction to $\bigcup_i M_i$ and its (unique) trace.

THEOREM. *The von Neumann algebra A_τ generated by all the $\mathscr{A}_{n,\tau}$ inside M_∞ is a II_1 factor.*

PROOF. It suffices to show that A_τ admits a unique normal trace (if the center were not trivial then one could construct other traces using elements of the center). But since $\bigcup_n \mathscr{A}_{n,\tau}$ is weakly dense in A_τ, it suffices to show, by Lemma 3.5, that any normal trace φ on A_τ agrees with tr on a product of commuting e_i's. To do this we use an infinite trick: given a commuting family of e_i's, extend it to an infinite family, which might as well be $e_2, e_4, e_6, \ldots$ for the rest of the argument. Then by (iv) of §3.5 we have $\mathrm{tr}(e_2 e_4 e_6 \cdots e_{2n}) = \tau^n$, etc., so that the von Neumann algebra generated by the e_{2n}'s is actually a "Bernoulli" algebra $\bigotimes_{i=1}^\infty (\mathbb{C} \oplus \mathbb{C})$ with the product trace corresponding to a "probability of success" of τ. I claim that any finite permutation of $\{e_2, e_4, \ldots\}$ can be effected by conjugation by a unitary in the normalizer of the algebra they generate. For this it suffices to show that e_2 and e_4 can be interchanged by a unitary in the algebra generated by $\{1, e_2, e_3, e_4\}$. Such a unitary can be constructed from 1 and $e_2 e_3 e_4$ since $(e_2 e_3 e_4)(e_4 e_3 e_2) = \tau^2 e_2$ and $(e_4 e_3 e_4)(e_2 e_3 e_4) = \tau^2 e_4$. (If there are gaps > 2 in the commuting sequence just use a longer descending product of e_i's.)

Thus our trace φ, when restricted to the Bernoulli algebra, defines a measure invariant under the obvious action of the infinite symmetric group S_∞, and absolutely continuous (by normality) with respect to the measure defined by tr. The action of S_∞ is easily seen to be ergodic, hence $\varphi = \mathrm{tr}$.

3.7. The element $e_1 \vee e_2 \vee \cdots \vee e_n$, the values $4\cos^2 \frac{\pi}{n}$. Let us define the polynomials $P_m(x)$ for $m = 0, 1, 2, \ldots$ by $P_0(x) = 0$, $P_1(x) = 1$, and $P_{m+1} = P_m - xP_{m-1}$, etc. By solving the difference equation we find easily that
$$P_n\left(\frac{1}{4\cos^2\theta}\right) = \frac{\sin n\theta}{2^{n-1}\cos^{n-1}\theta \sin\theta},$$
which makes the following assertions easy to prove. (That they are true is readily seen by drawing the graphs of the first few P_n's as in Figure 3.1.)

a) $P_n(\tau) > 0$ for $\tau < (4\cos^2 \frac{\pi}{n})^{-1}$
b) $P_{n+1}(\tau) < 0$ for $\frac{1}{4\cos^2 \pi/n+1} < \tau < \frac{1}{4\cos^2 \pi/n}$.

Projections in a von Neumann algebra form a complete lattice and we now want to calculate $e_1 \vee e_2 \vee e_3 \vee \cdots \vee e_n$ for each n. A purely algebraic way to describe this element is as the identity (it has one because it is a finite-dimensional von Neumann algebra) of the algebra generated by $e_1, e_2, \ldots, e_m$. We will see that the difference between the discrete and continuous series is the existence of a finite n for which $e_1 \vee e_2 \vee \cdots \vee e_n = 1$. The following beautiful formula for this supremum, and its proof, are due to Wenzl ([**We1**]). It is more convenient to define $f_n = 1 - e_1 \vee e_2 \vee \cdots \vee e_n$.

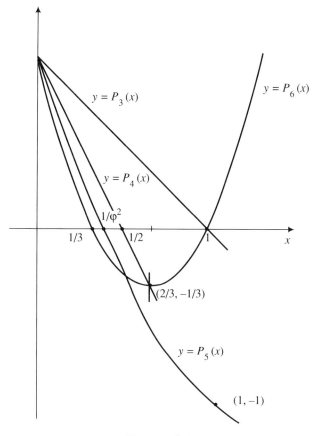

FIGURE 3.1

THEOREM. *If $N \subseteq M$ has index τ^{-1} and the e_i's in M_∞ are as usual, then if $P_k(\tau) > 0$ for $k \leq n + 2$ then $f_{n+1} = f_n - \frac{P_{n+1}(\tau)}{P_{n+2}(\tau)} f_n e_{n+1} f_n$.*

PROOF. Using induction and relations (i), (ii), and (iii) of §3.4, we can easily check that $(e_{n+1} f_n)^2 = (P_{n+2}/P_{n+1}) e_{n+1} f_n$ so that $(f_n e_{n+1} f_n)^2 = (P_{n+2}/P_{n+1}) f_n e_{n+1} f_n$ and the expression $f_n - (P_{n+1}/P_{n+2}) f_n e_{n+1} f_n$ is an orthogonal projection. It also follows that $e_{n+1} f_{n+1} = 0$ and $e_k f_{n+1} = 0$ for $k \leq n$ by induction. Thus $(1 - f_{n+1}) \geq \bigvee_{i=1}^{n+1} e_i$. But $1 - f_{n+1}$ is easily seen by induction to be in the algebra generated by $e_1, e_2, \ldots, e_n$ (without 1). Thus $1 - f_{n+1} = \bigvee_{i=1}^{n+1} e_i$.

With this result in hand it is easy to prove (a) of §3.1. The moral of the proof is that the sesquilinear form defined by $\langle a, b \rangle = \text{tr}(a^* b)$ has to be positive definite and we will show that if $\tau > \frac{1}{4}$ but different from one of the allowed values, then there are vectors of negative length for this form. Lest anyone think I am artificially making the proof as similar to §3.8 as possible, let me say that I distinctly remember diagonalizing the 14×14 matrix of inner products between the words on the e_i's in $\mathscr{A}_{3,\tau}$ of §3.6, in an early attempt to deduce restrictions on the index! I must have made a mistake as I did not get the restriction we will now obtain.

For if $\tau > \frac{1}{4}$ but not $1/(4\cos^2\frac{\pi}{n})$ for any n, then there are well-defined n and $n+1$ for which $1/(4\cos^2\pi/(n+1)) < \tau < 1/(4\cos^2\pi/n)$. From the properties of P_n it follows that $P_k(\tau) > 0$ for $k \leq n$ so that $f_{n-1} = f_{n-2} - (P_{n+1}/P_n)f_{n-2}e_{n-1}f_{n-2}$. Taking the trace using (iv) of §3.4 we see that

$$\text{tr}(f_{n-1}^*f_{n-1}) = \frac{P_n - \tau P_{n-1}}{P_n}\text{tr}(f_{n-2}) = \frac{P_{n+1}(\tau)}{P_n(\tau)}\text{tr}(f_{n-2}) < 0$$

since τ is in the appropriate interval. This is a contradiction. It is easy to check that if $\tau^{-1} = 4\cos^2\frac{\pi}{n}$ then $e_1 \vee e_2 \vee \cdots \vee e_{n-2} = 1$ and the formula of the theorem fails to work for larger values of n.

The formula of Wenzl is very useful for working with the algebra whose presentation is (i), (ii), (iii) of §3.4.

3.8. Ghosts. We now return to the Virasoro unitarity theorem of §3.2. To warm up let us prove that $c \geq 0$, $h \geq 0$. For consider $\|L_{-n}|h\rangle\|^2 = \langle h|L_n L_{-n}|h\rangle = \langle h|2nL_0 + c\frac{n^3-n}{12}|h\rangle = 2nh + c\frac{n^3-n}{12} \geq 0$. Putting $n = 1$, we get $h \geq 0$ and letting n grow we see $c \geq 0$.

The idea of the rest of the proof is just an elaboration of the above by considering the most general inner products. The dense subspace V is a direct sum of eigenspaces of L_0 which are orthogonal since $L_0^* = L_0$. For instance the $h+2$ eigenspace is spanned by $L_{-1}^2|h\rangle$ and $L_{-2}|h\rangle$ and we may quickly work out the inner products, $\|L_{-2}|h\rangle\|^2 = 4h + \frac{c}{2}$ (as above), $\|L_{-1}^2|h\rangle\|^2 = \langle h|L_1 L_1 L_{-1} L_{-1}|h\rangle = 2\langle h|L_1 L_0 L_{-1}|h\rangle + \langle h|L_1 L_{-1} L_1 L_{-1}|h\rangle = \cdots = 8h^2 + 4h$, and finally $\langle h|L_1^2 L_{-2}|h\rangle = 3\langle h|L_1 L_{-1}|h\rangle = 6h$ so that the matrix of inner products for these vectors is

$$\begin{pmatrix} 4h + \frac{c}{2} & 6h \\ 6h & 8h^2 + 4h \end{pmatrix}.$$

Since these are vectors in a Hilbert space, the eigenvalues must be nonnegative; so since $h \geq 0$, this is guaranteed by $\det \geq 0$, i.e., $16h^2 + (2c - 10)h + c \geq 0$.

This excludes a region on one side of a parabola in the (c, h)-plane. The reader will see that the analysis could be continued in principle to obtain more and more restrictions on c and h but without a nice formula for the eigenvalues of the matrices of inner products the general case would be painful. We invite the reader to try the $h + 3$ eigenspace of L_0.

Fortunately Kac in [**Ka1**] worked out the *determinant* of the matrix of inner products for the natural vectors in the $(h+n)$-eigenspace. Note that we need to consider only as many vectors as there are ways of writing n as a sum of positive integers, $p(n)$ (as far as the linear span is concerned the order of the L_i's doesn't count). Kac's formula is

$$\det{}_n = (\text{const}) \prod_{k=1}^{n} \psi_k(c, h)^{p(n-k)}$$

where

$$\psi_k(c, h) = \prod_{pq=k} (h - h_{p,q}(c)) \quad (p, q \text{ positive integers})$$

and

$$h_{p,q}(c) = \frac{[(m+1)p - mq]^2 - 1}{4m(m+1)},$$

m being a number related to c by $c = 1 - \frac{6}{m(m+1)}$. A careful analysis reveals that the only values for which this determinant is never negative (negativity would imply the contradictory existence of a "ghost" vector of negative length) are those allowed by the theorem.

As for the case $c \geq 1$, we see that the determinant is never negative for $h \geq 0$. Thus the inner product defined by the unitarity condition gives a pre-Hilbert space.

Of course we have not proven the existence of unitary representations of Vir realizing the values of c and h of Theorem 3.2. We will show how this is done in §4.4.

The analogy of the proof outlined above with that of §3.7 is clear but perhaps not surprising. Looking for negative length vectors is quite standard in deciding when a representation is unitary. The analogy is strengthened by considering the continuous series constructed in this section and the first construction in §3.3. The point is that nothing happens when $c \geq 1$; there is no collapsing and ultimately these representations are, by themselves, rather boring. In §3.3 we saw that the continuous variation of the index ≥ 4 was just due to the continuous dimensionality of projections in the relative commutant, which is quite straightforward compared with index phenomena.

Lecture 4. Construction of Examples, Further Structure

4.1. The discrete series of subfactors. One of the great strokes of luck about the proof of (a), Theorem 3.1, is that it contains an idea which could not be missed for the construction of examples of subfactors in the discrete series. For if the corresponding e_i algebra existed, it would, by §3.6, define a II_1 factor and one could consider the subfactor generalized by $\{e_2, e_3, \ldots\}$. We refer to [**J3**] for the proof that the index of this subfactor is indeed $4\cos^2\frac{\pi}{n}$, although it can be deduced from techniques to be developed in this lecture. But so far the argument is circuitous—we need the subfactor to construct the e_i's. The way to surmount this difficulty is to go to nonfactors. Then we can even consider finite-dimensional algebras, whose index we do not care about, to form towers in which the e_i's will live.

In our discussion of the basic construction in finite dimensions, we began with an arbitrary faithful trace on M and formed $\langle M, e\rangle$, e being the conditional expectation onto N. We could choose any faithful trace on $\langle M, e\rangle$ to redo the construction, and so on to form a tower. We would get a sequence of projections but there would be several problems. First, if the traces are not carefully chosen they will not agree with the previously chosen traces in the towers; so there will be no natural von Neumann algebra completion. Second, the relations (ii) and (iv) of §3.1 have no meaning since there is no parameter τ.

Fortunately, these two difficulties are quite compatible. If the Bratteli diagram for $N \subseteq M$ is connected, it is not hard to show that there is a unique (positive normalized) trace on any inductive limit algebra having the same Bratteli diagram as our tower would have, so that this trace may be used for all the basic constructions in the tower. Moreover, this trace defines a number τ for which the e_i's satisfy all the properties of §3.1.

We clearly need to understand traces on finite-dimensional von Neumann algebras. The picture is very simple since the trace on an $n \times n$ matrix algebra is unique up to a scalar multiple. Thus to define a (positive normalized) trace on the algebra defined by 3 2 5, we need to give three positive real numbers x_1, x_2, x_3 with $3x_1 + 2x_2 + 5x_3 = 1$. The x_i's will be the traces of minimal projections in the corresponding algebras. Thus a finite-dimensional von Neumann algebra with a positive normalized trace is defined by two vectors, **D** (= (3, 2, 5) in the above case) and **T** (= (x_1, x_2, x_3) in the above case) with **D** · **T** = 1.

It is high time we introduced the matrix Λ describing an inclusion $\mathbf{N} \subseteq \mathbf{M}$. If the center of **M** is m-dimensional and that of **N** is n-dimensional, Λ is the $m \times n$ matrix whose (i, j) entry is the number of lines on the Bratteli diagram between the ith number of M and the jth number of N. If **M**, defined by **D**, has a trace given by **T** whose restriction to **N** is given by **S**, and **N** is defined by **E**, one has obviously $\mathbf{D} = \Lambda \mathbf{E}$ and $\mathbf{S} = \Lambda^t \mathbf{T}$. Thus here is a diagramatic example of an inclusion with compatible traces, where we agree to put the weights of the traces (the x_1, x_2, x_3 as above) in circles:

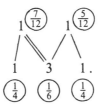

The result of use to us is the following.

THEOREM. *Suppose $N \subseteq M$ is an inclusion of finite-dimensional von Neumann algebras with connected Bratteli diagram. There is a unique trace, tr, on M, and a unique $\tau > 0$, with the property that, if one performs the basic construction of §2.6 with respect to tr, then it extends to a trace Tr on $\langle M, e \rangle$ such that the "Markov" condition, $\text{Tr}(xe) = \tau \text{Tr}(x)$ for $x \in M$, holds.*

Moreover, the vector defining Tr on M is an eigenvector of $\Lambda^t \Lambda$ belonging to the Perron eigenvalue (largest eigenvalue, of multiplicity one).

Finally, Tr has all the above properties for the pair $M \subseteq \langle M, e_N \rangle$.

This long-winded result is painful to prove rather than difficult. It is done in [**J3**]; see also [**GHJ**]. However, it is just what is needed to be able to iterate the tower construction and obtain e_i's with the right properties. The existence of subfactors of index $4\cos^2 \pi/n$ then follows from the existence of appropriate matrices of nonnegative integers. To obtain $4\cos^2 \pi/n$ start with the Bratteli diagram $1_{\searrow}2_{\searrow}1_{\searrow}2...2_{\searrow}1$ (or $1_{\searrow}2_{\searrow}1_{\searrow}2...2_{\searrow}1$) where there are $n - 1$ 1's and 2's altogether.

One can go further to find all possible Bratteli diagrams that can be used as seeds in the above construction. The situation is worked out completely in Chapter 1 of [**GHJ**]. The actual numbers are irrelevant; so we just give the answer in terms of the underlying graph of the Bratteli diagram. The possibilities are:

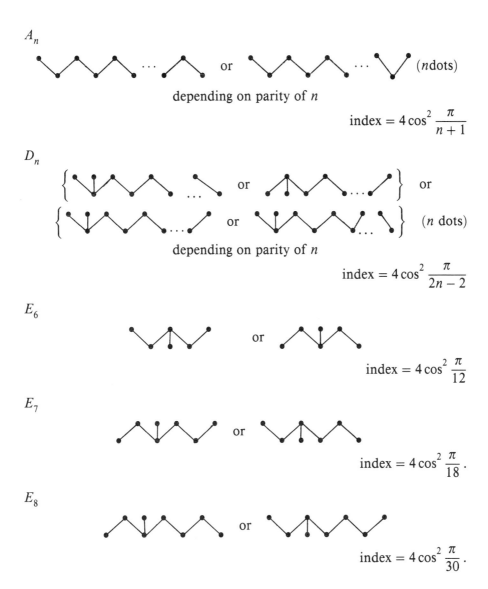

The occurrence of the A, D, E Dynkin diagrams is interesting. We will see them occur canonically in §4.6, and we note that this further deepens the analogy with conformal field theory where they occur in theories whose Virasoro algebra has central charge < 1 (see [**CIZ**]).

36 CONSTRUCTION OF EXAMPLES, FURTHER STRUCTURE

4.2. Bratteli diagrams of the e_i algebras. It is irresistible at this stage to give the Bratteli diagrams for the e_i algebras, according to the scheme $\mathscr{A}_{1,\tau} \subseteq \mathscr{A}_{2,\tau} \subseteq \mathscr{A}_{3,\tau} \subseteq \cdots$. We shall also record the restrictions to $\mathscr{A}_{n,\tau}$ of the trace according to the conventions of §6.1. It is convenient to start with the "generic" case $\tau \leq \frac{1}{4}$. It is remarkable that this diagram occurs as such in the thesis of A. Wassermann.

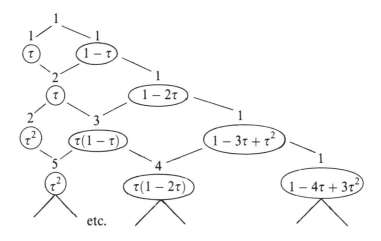

The diagram is like a truncated Pascal's triangle. The polynomials in τ appearing on the right-hand diagonal are the P_n's of §3.7. The weights for the trace differ simply by a multiplicative factor of τ down vertical lines.

It is possible to derive this structure by identifying the e_i algebra with a Hecke algebra quotient as in §9.2, and then use Iwahori's solution of the Hecke algebra structure. But it can also be done in a satisfactory way by invoking the rules for the basic construction. For we see from the diagram that $\mathscr{A}_{n+1,\tau}$ is related to the pair $\mathscr{A}_{n-1,\tau} \leq \mathscr{A}_{n,\tau}$ by a basic construction followed by addition of a single 1. This observation is used as a proof by applying the basic construction to the pair $\mathscr{A}_{n-1,\tau} \leq \mathscr{A}_{n,\tau}$ for the appropriate trace and observing that the e_i's in $\mathscr{A}_{n,\tau}$, together with the e of the basic construction, satisfy (i), (ii), (iii) of §3.1. The only problem is with the trace and that is a little messy. Suffice it to say that in order for the trace to be correct (remember it is the trace that really defines the structure) one is forced to add the final 1, which represents $1 - e_1 \vee \cdots \vee e_{n+1}$. For details see [**J3**] or [**GHJ**].

One advantage of deriving the structure in the way just sketched is that it generalizes immediately to the cases $\tau = 1/(4\cos^2 \frac{\pi}{m})$. The result is as follows. Suppose τ is the smallest real zero of one of the polynomials on the diagonal of the figure, say on the kth row. Then the previous argument for the structure of $\mathscr{A}_{\tau,n}$ works right down to the $(k-1)$th row. But when we construct the kth row we find that, since $1 - e_1 \vee \cdots \vee e_k = 0$, the whole of $\mathscr{A}_{\beta,k}$ is given by the basic construction and we must not add the extra one. From then on the basic construction suffices to obtain $\mathscr{A}_{n+1,\tau}$ from $\mathscr{A}_{n-1,\tau} \subseteq \mathscr{A}_{n,\tau}$. We illustrate with the first few cases: $\tau^{-1} = 4\cos^2 \frac{\pi}{4} = 2$.

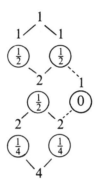

The alert reader will recognize the Clifford algebra!

$$\tau^{-1} = 4\cos^2 \tfrac{\pi}{5} = \varphi^2 \quad (1 - 3\tau + \tau^2 = 0)$$
(Fibonacci numbers, of course)

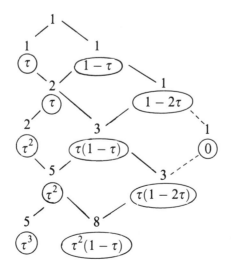

$$\tau^{-1} = 4\cos^2 \tfrac{\pi}{6} = 3 \quad (1 - 4\tau + 3\tau^2 = 0)$$

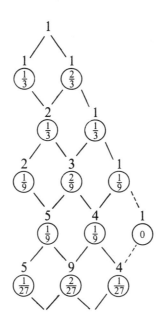

4.3. Affine Lie algebra. If G is a compact Lie group with complexified Lie algebra $\mathscr{G}$, one may consider groups LG of "loops", i.e., suitably smooth maps from S^1 to G under pointwise multiplication. These may be given an infinite-dimensional Lie group structure and, in the same spirit as §3.2, we consider the Lie algebra of polynomial loops $\mathscr{G} \otimes \mathbb{C}[t, t^{-1}]$ as a denatured version of the complexified Lie algebra of LG. The whole subject would be devoid of interest (for me) were it not for the existence of nontrivial central extensions of $\mathscr{G} \otimes \mathbb{C}[t, t^{-1}]$ (note that the Lie bracket of $\mathscr{G} \otimes \mathbb{C}[t, t^{-1}]$ is $[x \otimes f, y \otimes g] = [x, f] \otimes fg$). To simplify matters let us assume that $\mathscr{G}$ is simple; so there is an essentially unique invariant bilinear form $\langle\ ,\ \rangle$ on $\mathscr{G}$. One may then use this form to create a universal central extension of $\mathscr{G} \otimes \mathbb{C}[t, t^{-1}]$ sometimes called $\widehat{\mathscr{G}}$, namely $\mathscr{G} \otimes \mathbb{C}[t, t^{-1}] \oplus \mathbb{C}c$ with bracket $[c, x \otimes f] = 0$ and $[x \otimes t^n, y \otimes t^m] = [x, y] \otimes t^{n+m} + m\delta_{n,-m}\langle x, y\rangle c$.

There are good reasons for extending $\mathscr{G}$ even further by forming the semidirect product with a one-dimensional Lie algebra acting on $\mathscr{G}$, the derivation D defined by $Dc = 0$, $D(x \otimes t^n) = nx \otimes t^n$. Call the resulting Lie algebra $\widetilde{\mathscr{G}}$, i.e., $\widetilde{\mathscr{G}} = \widehat{\mathscr{G}} \oplus \mathbb{C}d$, $[d, a] = D(a)$ for $a \in \widehat{\mathscr{G}}$. One reason for introducing D is to have a nondegenerate invariant bilinear form on $\widetilde{\mathscr{G}}$ (impossible on $\widehat{\mathscr{G}}$) defined by $\langle x \otimes t^m, y \otimes t^n\rangle = \delta_{n,-m}\langle x, y\rangle$, $\langle x \otimes f, c\rangle = \langle x \otimes f, d\rangle = \langle c, c\rangle = \langle d, d\rangle = 0$, $\langle c, d\rangle = 1$. Another reason is that one then obtains a beautiful "root space" decomposition into one-dimensional eigenspaces for the adjoint representation of the abelian subalgebra $\mathcal{N} \otimes 1 \oplus \mathbb{C}c \oplus \mathbb{C}d$, $\mathcal{N}$ being a Cartan subalgebra of $\mathcal{G}$. A third reason is that one obtains precisely what is known as a Kac-Moody Lie algebra with defining presentation given simply in terms of the extended Cartan matrix of $\mathcal{G}$.

The whole picture can be complicated by starting, not from LG, but taking an automorphism α of G and taking functions $f\colon \mathbb{R} \to G$ with $f(\theta+2\pi) = \alpha(f(0))$.

You will hear all of the Lie algebras constructed above called affine Lie algebras at some time by someone. They are by far the most tractable and useful among the more general Kac-Moody Lie algebras. See [**Ka2**] and [**PS**].

4.4. Realizing the Virasoro discrete series. As we have seen with the e_i's, it is possible to construct the algebras with $1/\tau = 4\cos^2 \frac{\pi}{n}$ by hand, but it is rather painful. On the other hand, it becomes trivial by applying the tower theory of §3.5 to a relative nonfactor situation. So it is with the Virasoro discrete series, although the difficulties with a "by hand" construction seem more serious. The following beautiful trick for constructing the discrete series is due to Goddard, Kent, and Olive in [**GKO**].

The theory we need is that of the "Segal-Sugawara" form. Projective unitary representations of LG are acted on naturally by $\mathrm{Diff}\, S^1$ so that if some representation in $\mathcal{H}$ is isolated, we expect, on abstract grounds as in the Mackey induced representation machine, a projective unitary representation of $\mathrm{Diff}^+ S^1$ on $\mathcal{H}$, appropriately intertwining the projective representation of LG.

The Segal-Sugawara form gives an explicit formula for the action of the generators L_n, and thus the central charge, all worked on the denatured level of Vir and the affine Lie algebra, on some dense subspace of the Hilbert space.

The representations of LG we shall consider will be the "lowest weight" positive energy ones. That is, we will assume that all of $\mathcal{G}$ acts and that the Hilbert space is a direct sum of the (finite-dimensional) eigenspaces of d, only nonnegative values occurring as eigenvalues. Finite linear combinations of eigenvectors of α are called "finite energy vectors". Such representations may be constructed from certain geometrical Hilbert spaces and so are manifestly unitary (see [**PS**]).*

The key trivial remark to get started is that, in a positive energy representation, any finite energy vector is killed by all $x \otimes t^{-n}$ for n sufficiently large (use $[d, x \otimes t^{-n}] = -nx \otimes t^{-n}$). This is the guiding principle which allows all subsequent formulae to make sense.

So let us suppose we are given a positive energy representation of $\mathcal{G}$ for which c is a scalar multiple of the identity. Let V be the space of finite energy vectors. Choose an orthonormal basis e_a of $\mathcal{G}$. Let σ be the value of

---
* It's not really that easy!

the Casimir $\sum e_a^2$ in the adjoint representation of $\mathscr{G}$ and put $N = \dim \mathscr{G}$. On V define

$$: e_a^m e_a^n := \begin{cases} (e_a \otimes t^m)(e_a \otimes t^n) & \text{if } m \geq 0, \\ (e_a \otimes t^n)(e_a \otimes t^m) & \text{if } m < 0. \end{cases}$$

THEOREM. *For each* $m \in \lambda$ *the sum* $L_m = -\frac{1}{2} \sum_{k \in \mathbb{Z}} \sum_a : e_a^k e_a^{m-k} :$ *is well defined on* V *and*

a) $[L_m, e_a \otimes t^n] = n e_a \otimes t^{m+n}$,

b) $[L_m, L_n] = (n-m)L_{m+n} + \delta_{m,-n}(\frac{Nc}{c+\sigma})(\frac{m^3-m}{12})$.

The proof is a very enjoyable computation requiring a lot of care with potentially infinite sums.

Formula a) shows that the Virasoro action intertwines the action of $\widehat{\mathscr{G}}$ correctly and b), which follows from a), shows that it is indeed a Virasoro action. That $L_n^* = L_{-n}$ follows from the unitarity of the representation of $\widehat{\mathscr{G}}$.

A look at specific cases (c is an integer) shows that there is no chance of obtaining values of the Virasoro central charge less than one.

However, we may consider a relative situation where we start with $\mathscr{G}_0 \subseteq \mathscr{G}$ a subalgebra. An orthonormal basis e_a, $a \in A$, for $\mathscr{G}_0$ may be extended to an orthonormal basis e_b, $b \in B$, for $\mathscr{G}$. The affine Lie algebra $\widehat{\mathscr{G}}_0$ is naturally contained in $\widehat{\mathscr{G}}$ and a positive energy representation of $\widehat{\mathscr{G}}$ restricts to a positive energy representation of $\widehat{\mathscr{G}}_0$. Now let $L_m = \sum_{\substack{k \in \mathbb{Z} \\ b \in B}} : e_b^k e_b^{m-k} :$, $\mathscr{L}_m = \sum_{\substack{b \in \mathbb{Z} \\ a \in A}} : e_a^k e_a^{m-k} :$. Then for $a \in A$ we have, by a) of the theorem, $[L_m, e_a \otimes t^n] = n e_a \otimes t^{m+n}$ and $[\mathscr{L}_m, e_a \otimes t^n] = n e_a \otimes t^{m+n}$; so by the calculation that proves b) from a), $[L_m, L_n] = [\mathscr{L}_m, L_n]$ so that $L_m - \mathscr{L}_m$ commutes with L_n for all n and $L_m - \mathscr{L}_m$ gives a unitary Virasoro representation whose central charge is the difference between those of L_n and $\mathscr{L}_n$.

Looking at examples we now find all the values of c and n allowed for $c < 1$ by Theorem 3.2.

4.5. The tower of relative commutants. If we return to §3.5, we see that we have used nothing but the e_i's in the tower constructed from a pair $N \subseteq M$ of finite-dimensional von Neumann algebras. This is presumably analogous to using only the Virasoro algebra in a positive energy representation of an affine Lie algebra. There is much more there. Consider again the tower $N \subseteq M \subseteq M_2 \subseteq M_3 \subseteq M_4 \subseteq \cdots$. Since $[M_k : N] = [M : N]^k < \infty$ we have $\dim(N' \cap M_k) < \infty$ by (v) of §2.5. Similarly $M' \cap M_k$ is finite-dimensional and we have two canonical increasing sequences of finite-dimensional von Neumann algebras with canonical traces on them. They each contain the e_i's as a subfactor.

Much attention has been paid to the two towers by Ocneanu. He has shown that if a condition called finite depth (to be defined below) is satisfied by $N \subseteq M$, and M is hyperfinite, then the weak closures of the two towers of relative commutants in M_∞ give a subfactor which is isomorphic to the inclusion $N \subseteq M$ (or possibly $M \subseteq \langle M, e \rangle$, I'm not sure; see [**Oc2**] for the true story). This is a very powerful result since the pair inside the tower is completely explicitly constructible. In this way Ocneanu has classified all subfactors of R of index less than 4.[*]

We shall content ourselves with giving a qualitative description of the towers in terms of their Bratteli diagrams and working a couple of examples. This is an area of very active interest at the moment. For this qualitative description we consider only $N' \cap M_i$.

Recall that M_{i+1} is obtained from M_i by adding the projection e_i onto M_{i-1}. By (b) of (2.6), $e_i \in N' \cap M_{i+1}$. It is easy to check that $e_i(N' \cap M_i)e_i \subseteq N' \cap M_{i-1}$; indeed $e_i x e_i = E(x)e_i$ for $x \in N' \cap M_i$ where E is the conditional expectation onto $N' \cap M_{i-1}$ with respect to the ambient trace. This alone guarantees the existence of a copy of the basic construction for the pair $N' \cap M_{i-1} \subseteq N' \cap M_i$ inside $N' \cap M_{i+1}$. It may or may not be all of $N' \cap M_{i+1}$. By considering the trace one may show that, if there is an i for which this basic construction gives all of $N' \cap M_{i+1}$, then it is true for all $k \geq i$. Thus in this case we know the Bratteli diagram for the whole tower as soon as we know it for $N' \cap M_{i-1} \subseteq N' \cap M_i$, since we saw in §2.7 how to calculate the basic construction in finite dimensions. If an i as above exists we say $N \subseteq M$ is of *finite depth* and the depth is then the smallest such i. Thus in this case the tower might typically look like

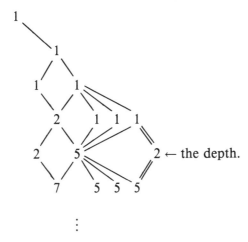

It is also possible that no such i exists. In that case the tower continues

---

[*] The main result for finite depth subfactors has been proved by Popa in [**Pop**].

to grow and an example might be Pascal's triangle:

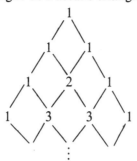

We shall see below examples of both kinds of behaviour. For details of the above see [**GHJ**].

Note that in the finite depth case, once the tower has stabilized, the inclusions are given by matrices, alternately Λ and Λ^t for some matrix Λ. It is part of the theory that $[M:N] = \|\Lambda^t\Lambda\|$—an algebraic integer (of a very special kind). In conformal field theory there is a notion analogous to finite depth which is known to imply rationality of the central charge [**AM**]. Moreover, it is conjectured that the central charge is rational in any conformal field theory. Similarly in subfactors it is (weakly) conjectured that, if the relative commutant $N' \cap M$ is trivial, $[M:N]$ is an algebraic integer!*

4.6. Examples of towers of relative commutants. The reader may have noticed that the Bratteli diagrams of §4.2 look like the kind that could occur. To show this we use the beautiful lemma of C. Skau which follows.

LEMMA. *Let $N \subseteq M \subseteq \langle M, e_1 \rangle \subsetneq \langle M, e_1, e_2 \rangle \subseteq \cdots \subseteq M_\infty$ be as usual (M_∞ a II_1 factor). Then if $[M:N] \le 4$, $\{e_1, e_2, \ldots\}' \cap M_\infty = N$.*

PROOF. Note first that it is obvious from §2.6(b) that if $x \in \langle M, e_1, \ldots, e_k \rangle \cap \{e_1, e_2, \ldots\}'$ then $x \in N$. Let $F_k: M_\infty \to \{e_k, e_{k+1}, \ldots\}' \cap M_\infty$ be the canonical conditional expectation. Since $F_1 F_k = F_1$, it suffices to show that $F_k(M_\infty) \subseteq M_{k-1}$. But F_k is weakly continuous; so it suffices to show $F_k(M_k) \subseteq M_{k-1}$ for every k. But by (d) of §2.6.2 it suffices to show that $F_k(x e_{k-1} y) \in M_{k-1}$ for $x, y \in M_{k-1}$. Now x and y are in $\{e_k, e_{k+1}, \ldots\}'$; so by the bimodule property, $F_k(x e_{k-1} y) = x F_k(e_{k-1}) y$. But by (vi) of §2.5, the subfactor $\{e_k, e_{k+1}, \ldots\}''$ of $\{e_{k-1}, e_k, \ldots\}''$ has trivial relative commutant which assures $F_k(e_{k-1})$ is a scalar.

Now we can see that, if $P_{4\cos^2(\pi/n)} \subseteq M$ are the subfactors $\{e_2, e_3, \ldots\}'' \subseteq \{e_1, e_2, e_3, \ldots\}''$ constructed in §4.1 then the tower of relative commutants is as in §4.2. For if the e_i's of the tower for $P_{4\cos^2(1/\pi)} \subseteq P$ are written f_i, then M_i is $\{\ldots, e_3, e_2, e_1, f_1, f_2, \ldots, f_i\}''$, and by Skau's lemma applied backwards, $P'_{4\cos^2(\pi/n)} \cap M_i$ is $\{f_1, f_2, \ldots, f_i\}''$.

---

* Now only in the hyperfinite case because of Popa's results.

In fact we know what the possibilities are in index < 4 for the stabilized matrices Λ, Λ^t (index < 4 implies finite depth). They are given as in §4.1 by the A, D, E Coxeter graphs. Ocneanu has shown ([Oc2]) that D_n is impossible for n odd, that E_7 is impossible and that for A_n and D_n there is a unique subfactor of R, up to conjugacy, with the Coxeter graph giving the stabilized Bratteli diagram. Also, there are exactly two different subfactors for each of E_6 and E_8, one obtained from the other by applying an antiautomorphism.

One might be tempted to think that the Bratteli diagram of the tower of relative commutants for the pair $N \subseteq M$ of index $r > 4$ is the same as the diagram for the e_i's at the beginning of §4.2. This is not the case as we shall see later. For the moment we content ourselves to observe that the tower of relative commutants for the pair of construction B of §3.3 is Pascal's triangle! This is easy to show since each step of the tower of M_i's is easily described by exploiting the simple type III picture. Indeed, if N is the centralizer of φ_λ on $1 \otimes \bigotimes_{i=1}^\infty M_2(\mathbb{C})$ and M is the centralizer of φ_λ on $M_2(\mathbb{C}) \otimes \bigotimes_{i=1}^\infty M_2(\mathbb{C})$ then $\langle M, e \rangle$ is the centralizer of φ_λ on $M_2(\mathbb{C}) \otimes M = M_2(\mathbb{C}) \otimes M_2(\mathbb{C}) \otimes (\bigotimes_{i=1}^\infty M_2(\mathbb{C}))$, and so on. Now one can use the fact that S_∞, acting in a natural way on $\bigotimes_{i=1}^\infty \mathbb{C}^2$, is in N. This already forces $N' \cap (\bigotimes_{i=1}^k M_2(\mathbb{C}) \otimes \bigotimes_{i=1}^\infty M_2(\mathbb{C}))$ to be contained in $\bigotimes_{i=1}^\infty M_2(\mathbb{C}) \otimes 1$. Thus the tower of relative commutants is the same as the tower of algebras given by the centralizers of the φ_λ's on $\bigotimes_{i=1}^k M_2(\mathbb{C})$. This is an easy finite-dimensional calculation that yields Pascal's triangle quite easily.

(Note that we have been quite abusive about using the same notation φ_λ for Power states on different von Neumann algebras.)

4.7. The relative commutant problem. As we have often hinted, one of the main remaining problems in subfactors is: What is the set of index values of subfactors N of R with $N' \cap R = \mathbb{C}$?

We know that this set is closed under multiplication, contains all integers ≥ 1, contains $4\cos^2 \frac{\pi}{n}$, and contains many other sequences of algebraic integers constructed by Wenzl ([We2], and §9.4). There seems to be a deal of mystery here and it is my hope that new constructions will emerge by pursuing the analogies with conformal field theory.

In the meantime let us give the example that gives the smallest known (at the time of writing)[*] value of this set, > 4. It is $3 + \sqrt{3}$, a subfactor for which is given as follows.

---

[*] Lower values have since been found by Haagerup and Schou, and Ocneanu.

Let $A \subseteq B$ be finite-dimensional von Neumann algebras with Bratteli diagram

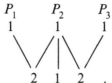

Let P_1, P_2, P_3 be the minimal projectors in A as indicated above. We saw in §4.1 how to find a trace on B which allows one to iterate the tower construction to get $A \subseteq B \subseteq \langle B, e_1 \rangle \subseteq \langle B, e_1, e_2 \rangle \subseteq \cdots \subseteq B_\infty$ with the e_i's satisfying (i), (ii), (iii), (iv) of §3.4 with $\tau^{-1} = 4\cos^2 \frac{\pi}{12}$. Do this. Then by Skau's Lemma 4.7, $\{e_1, e_2, e_3, \ldots\}' \cap B_\infty$ is just A. So the subfactor $P_1 \{e_1, e_2, e_3, \ldots\}'' \subseteq P_1 B_\infty P_1$ has trivial relative commutant. The whole tower $P\{e_1, e_2, \ldots, e_k\}'' \subseteq P_1 \{B, e_1, \ldots, e_k\}'' P_1$ is finite-dimensional and readily calculable. It obviously stabilizes at a certain point and one can calculate all the inclusion matrices. Invoking results of Wenzl [**We2**] or Pimsner and Popa, [**PP1**], one may calculate the index of the resulting subfactor and one obtains $3 + \sqrt{3}$. This procedure can obviously be repeated with different Coxeter graphs and different minimal central projections to obtain different indices of subfactors with trivial relative commutant. This is all worked out in detail in [**GHJ**].

Lecture 5. The Braid Group and Its Representations

5.1. Definition and presentation. The n-string braid group B_n is obtained by fixing n points in $\mathbb{R}^2$ and the same n points in a parallel copy of $\mathbb{R}^2$ underneath the first. A braid is then obtained by attaching the top n points to the bottom n by strings in such a way that the strings never intersect and such that the tangent vector to a string is never parallel to the planes. It is useful to arrange the n points on a straight line; so we may draw pictures of braids like the 4-string braid below:

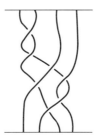

Obviously we will only consider braids up to some equivalence relation which will be defined precisely in our second formulation of the braid group below but may be considered to be the "obvious" relation: one may pull the strings around arbitrarily provided they do not cross and the tangent vectors never become parallel to the planes. See [**Bi1**].

The really useful thing about braids is that they form groups under the obvious concatenation; thus

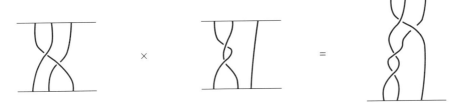

Associativity of the group law is obvious. The braid whose strings are perpendicular to the two planes is obviously an identity, and the mirror image of a braid in the bottom plane is its inverse.

The alternative and important approach to braids is to see them as elements of a certain fundamental group. Thus if we form $\mathbb{R}^2 \times \mathbb{R}^2 \times \cdots \times \mathbb{R}^2$ (n copies), we can think of our n points as defining an element $\mathbf{x}$ of this Cartesian product. In fact one may remove the "diagonal" $\Delta = \{(x_1, x_2, \ldots, x_n) | \exists i \neq j \text{ with } x_i = x_j\}$ and the points are in $\mathbb{R}^{2n} \setminus \Delta$. If we choose an element of $\pi_1(\mathbb{R}^{2n} \setminus \Delta; \mathbf{x})$ represented by a path $c(t)$, then $c(t)$ will define n distinct points in a plane t units below the top plane. Thus we will get a braid as above. But all such braids will not permute the set of n points defining $\mathbf{x}$. In order to obtain all braids consider instead $(\mathbb{R}^{2n} \setminus \Delta)/S_n$, S_n being the symmetric group acting in the obvious way. Then it is clear that $\pi_1((\mathbb{R}^{2n} \setminus \Delta)/S_n)$ can serve as a definition for B_n, since the homotopy equivalence of paths corresponds well to the intuitive equivalence explained above. The space $(\mathbb{R}^{2n} \setminus \Delta)/S_n$ is called "configuration space" and is rather important, especially if we identify $\mathbb{R}^2$ with $\mathbb{C}$. Configuration space is then, for instance, the space of all regular elements in a Cartan subalgebra of $\mathrm{GL}_n(\mathbb{C})$, divided by the action of the Weyl group. Or it is the space of (monic) polynomials of degree n over $\mathbb{C}$ with n distinct roots. We can thus expect the braid group to appear often in mathematics.

The braid group may be converted into a purely algebraic object by giving a presentation. This was first found by E. Artin [**Ar**]. The usual presentation of B_n begins with braids as pictured geometrically above and takes as generators $\sigma_1, \sigma_2, \ldots, \sigma_{n-1}$ where σ_i is as below:

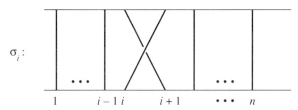

The relations $\sigma_i \sigma_j = \sigma_j \sigma_i$ if $|i-j| \leq 1$ and $\sigma_i \sigma_{i+1} \sigma_i = \sigma_{i+1} \sigma_i \sigma_{i+1}$ are proved by drawing pictures. Artin's theorem says that these generators and relations are a *presentation* of the group B_n. This is a very useful result as it allows us to construct representations of B_n simply by finding invertible matrices that satisfy these relations.

As an abstract group the braid group is torsion free and has center $\cong \mathbb{Z}$, a generator (for $n \geq 2$) of the center being $(\sigma_1 \sigma_2 \cdots \sigma_{n-1})^n$. This element actually has a square root Δ of some importance. This Δ is a half twist algebraically written $(\sigma_1 \sigma_2 \cdots \sigma_{n-1})(\sigma_1 \sigma_2 \cdots \sigma_{n-2}) \cdots (\sigma_1 \sigma_2) \sigma_1$. There is a kind of "polar decomposition" of a braid word: any braid word can be written $\Delta^k P$ where P is a word on $\sigma_1, \sigma_2, \ldots, \sigma_{n+1}$ (no inverses). Such a decomposition can be made unique and is the basis of Garside's solution [**Ga**] of the word and conjugacy problems in B_n.

5.2. Action of the braid group on the free group.
The group B_n acts in an obvious geometrical way on the free group on n generators as follows: every braid can be thought of as an isotopic family of embeddings of the set of n points into the disc D^2. By the isotopy extension theorem, this can be extended to an isotopy of diffeomorphisms of D^2 which are the identity on the boundary (and near it if necessary). Choosing a base point near the boundary we see that B_n acts on $\pi_1(D^2 - \{n \text{ points}\}) = F_n$. The action can be "seen" by making the braid out of wire and an element of $\pi_1(D^2 - \{n \text{ points}\})$ by an elastic string in the top plane of the braid. Forcing the string down to the bottom of the braid gives another element of $\pi_1(D^2 - \{n \text{ points}\})$; this is the image of the original one under the action.

The action is easy to describe in terms of generators. If the n points and base point are as below

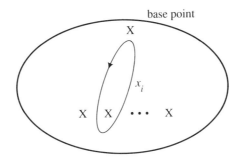

and x_i are generators of the fundamental group as drawn, then

$$\sigma_i(x_j) = \begin{cases} x_j, & \text{if } j \notin \{i, i+1\}, \\ x_i x_{i+1} x_i^{-1} & \text{if } j = i, \\ x_i & \text{if } j = i+1. \end{cases}$$

This representation of B_n in $\operatorname{Aut}(F_n)$ is very natural. It was shown by Artin to be faithful. Automorphisms α of F_n that are in the image of this representation are characterized by the following two properties:

(i) $\alpha(x_i)$ is a conjugate of some x_j,
(ii) $\alpha(x_1 x_2 \cdots x_n) = x_1 x_2 \cdots x_n$.

5.3. The pure braid group and the inductive structure of the braid groups.
In fact, the group B_n is not very complicated as a discrete group. B_2 is just $\mathbb{Z}$ and $B_3/Z(B_3)$ is easily seen to be $\operatorname{PSL}_2(\mathbb{Z})$, B_3 itself being the inverse image in $\widetilde{\operatorname{PSL}_2}(\mathbb{R})$ of $\operatorname{PSL}_2(\mathbb{Z})$. But for higher n it is convenient to study the normal subgroup P_n of pure braids, namely those which induce the trivial permutation on the end points of the braid. Obviously B_n/P_n is the symmetric group S_n. Looking at the above figure and moving the base point

into line with the other points we see that the group P_n contains a copy of the free group F_{n-1}. In terms of the σ_i's, the generators are $x_1 = \sigma_1^2$, $x_2 = \sigma_1 \sigma_2^2 \sigma_1^{-1}$, $x_3 = \sigma_1 \sigma_2 \sigma_3^2 \sigma_2^{-1} \sigma_1^{-1}$, and so on up to x_{n-1}. The nice thing about this subgroup is that it is a normal subgroup of P_n and by a procedure known as "combing" a braid, Artin showed that P_n is actually a semidirect product of F_{n-1} and P_{n-1}. Thus P_n is an iterated semidirect product of free groups, and B_n is an extension of finite index of P_n, a fact that is useful for calculating such things as cohomology rings.

5.4. Burau and Gassner representations. Being a group, B_n should be studied for its representations. The most natural and accessible of these is the Burau representation, which can be easily derived from the action of the braid group on the free group as follows. Let $\varepsilon\colon F_n \to \mathbb{Z}$ be the homomorphism that sends each x_i to 1. Obviously, $\varepsilon(\alpha(x)) = \varepsilon(x)$ if α is a braid group element acting on $x \in F_n$. Thus $\ker \varepsilon_n$ is invariant under the action of B_n. But $\ker \varepsilon_n$ is an easily understood subgroup, and one sees $\ker \varepsilon_n / [\ker \varepsilon_n, \ker \varepsilon_n]$ is a free abelian group of infinite rank but free of rank $n - 1$ as a $\mathbb{Z}[t, t^{-1}]$ module, "t" being the generator of the natural $\mathbb{Z}$-action. A basis is given by the classes modulo $[\ker \varepsilon_n, \ker \varepsilon_n]$ of $x_1^{-1} x_i$, $i = 2, 3, \ldots, n$. The action of B_n is clearly $\mathbb{Z}[t, t^{-1}]$-linear; so we deduce a representation of B_n in the $n-1 \times n-1$ matrices over $\mathbb{Z}[t, t^{-1}]$.

Let us calculate the matrix of σ_i with respect to the basis $(x_1^{-1} x_2)$, $(x_2^{-1} x_3)$, $\ldots$, $(x_{n-1}^{-1} x_n)$ of $\ker \varepsilon / [\ker \varepsilon, \ker \varepsilon]$:

a) $\sigma_1(x_1^{-1} x_2) = x_1 x_2^{-1} x_1^{-1} x_1 = x_1 x_2^{-1} = x_1 (x_2^{-1} x_1) x_1^{-1}$, but conjugation by x_1 induces "t" on the abelianization, so $\sigma_1[x_1^{-1} x_2] = -t[x_1^{-1} x_2]$;

b) $\sigma_1(x_2^{-1} x_3) = x_1^{-1} x_3 = (x_1^{-1} x_2)(x_2^{-1} x_3)$, so $\sigma_1[x_2^{-1} x_3] = [x_1^{-1} x_2] + [x_2^{-1} x_3]$;

c) $\sigma_2(x_1^{-1} x_2) = x_1^{-1}(x_2 x_3 x_2^{-1}) = (x_1^{-1} x_2) x_3 (x_2^{-1} x_3) x_3^{-1}$, so as in (a), $\sigma_2[x_1^{-1} x_2] = [x_1^{-1} x_2] + t[x_2^{-1} x_3]$.

Changing the indices we see that, as matrices, we have:

$$\sigma_1 \mapsto \begin{pmatrix} -t & 0 & 0 & & \\ 1 & 1 & 0 & & \\ & & 1 & & \\ 0 & & & \ddots & \\ 0 & & & & 1 \end{pmatrix}, \quad \sigma_i \mapsto \begin{pmatrix} 1 & & & & & \\ & \ddots & & & & \\ & & 1 & t & 0 & \\ & & 0 & -t & 0 & \\ & & 0 & 1 & 1 & \\ & & & & & \ddots \\ & & & & & & 1 \end{pmatrix},$$

$$\sigma_{n-1} \mapsto \begin{pmatrix} 1 & & & & & & \\ & 1 & & & & & \\ & & 1 & & & & \\ & & & 1 & & & \\ & & & & \ddots & & \\ & & & & & 1 & t \\ & & & & & 0 & -t \end{pmatrix}.$$

This representation is known as the Burau or "reduced Burau" representation. The last name is because one could have discovered the representation by reducing an even simpler n-dimensional one given in terms of generators and relations by

$$\sigma_i \to \begin{pmatrix} 1 & & & & & & & \\ & 1 & & & & 0 & & \\ & & \ddots & & & & & \\ & & & 1-t & 1 & & & \\ & & & t & 0 & & & \\ & & 0 & & & 1 & & \\ & & & & & & \ddots & \\ & & & & & & & 1 \end{pmatrix}$$

with the "$1 - t$" in the (i, i)-entry. (Reducibility follows from the fact that these are column-stochastic matrices.)

If one replaces the homomorphism ε by the abelianization map ab to $\mathbb{Z}^n$, only the pure braid group P_n acts on $\ker ab / [\ker ab, \ker ab]$ as $\mathbb{Z}[t_1, t_2, \ldots, t_n]$-linear maps. The ensuing representation of the pure braid group is called the Gassner representation.

It is a rather famous open problem to decide whether the Burau and/or Gassner representations are faithful. Using standard facts about $\mathrm{SL}_2(\mathbb{Z})$, it is easy to show that the Burau representation of B_3 is faithful. The question remains often for B_4 and P_4.*

---

*But Moody has recently found a 9-string braid in the kernel of the Burau representation!

Another useful approach to the Burau representation is to think of it as a "t-deformation" of the basic representation associated with the A_{n-1} Coxeter graph (see [**Bo**]). Thus one considers a free $\mathbb{Z}[s, s^{-1}]$ module with basis $v_1, v_2, \ldots, v_{n-1}$ and sesquilinear form defined by $\langle v_i, v_i \rangle = s + s^{-1}$, $\langle v_i, v_{i+1}\rangle = -1$. Then σ_i, for $i = 1, 2, \ldots, n-1$, is the pseudo-reflection $\sigma_i(v) = v - s\langle e_i, v\rangle e_i$. The braid relations are immediate as is the unitary of the representation as noted by Squier in [**Sq**]. See also [**GJ**].

5.5. Representations in the e_i algebras. The Burau and Gassner representations were very naturally obtained from the action of the braid group on the free group. Indeed, there is a purely geometric way to obtain these representations in terms of covering spaces (left as an exercise). The representations of most interest to us will be obtained in a far less satisfactory way—by the generators and relations picture. This lack of naturality is at the heart of our current lack of geometric insight into the link invariants to be defined later.*

The idea is quite simple. Try to send σ_i to $ae_i + b1$ where e_i are as in §3.4 and see if it is possible to choose a and b so that (i), (ii), and (iii) imply $\sigma_i \sigma_{i+1} \sigma_i = \sigma_{i+1} \sigma_i \sigma_{i+1}$. The calculation is simple and one finds a solution which is unique up to a scalar multiple. It is $a = (t+1)$, $b = -1$, where $\tau^{-1} = 2 + t + t^{-1}$.

Thus we have constructed for all n, a representation of B_n inside the e_i algebra, with a parameter t. The first thing to check is its relation with the Burau representation. For this, suppose we are in the continuous series so that we know the structure of the e_i algebra is given by

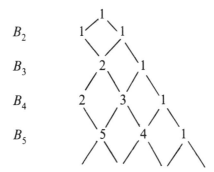

The interpretation of the diagonal lines is now the restriction of representations: the representation of the braid group B_5 corresponding to the "5" restricts to a direct sum of the representations corresponding to the "2" and "3". One can show that, up to sign, the Burau representation of B_n is equivalent to the representation of dimension $n - 1$ on the corresponding row

---

*Ruth Lawrence and others have obtained these representations from natural actions of the braid groups on homology groups.

of the diagram. There are several ways to prove this, the simplest probably being to exhibit the Burau matrices for the σ_i's in the form $ae_i + b$ and show that the relations (i), (ii), (iii) of §3.4 are satisfied (see [**J4**]).

It is perhaps more interesting to see what happens for the e_i-discrete series, when $\tau^{-1} = 4\cos^2\frac{\pi}{k}$ for some $k \in \mathbb{Z}$, $k \geq 3$. It may be shown that the image of the braid group is finite (for all n) when $\tau = 1/2$, $k = 4$. It is infinite for $k = 5$, but the most interesting phenomenon occurs when $k = 6$. Here the braid group representation once again factors through a finite group. The situation is analyzed and generalized in [**GJ**]. The thing that is going on is the Shale-Weil or metaplectic representation of the symplectic group $\text{Sp}(2n, \mathbb{Z}/3\mathbb{Z})$. The Bratteli diagram is

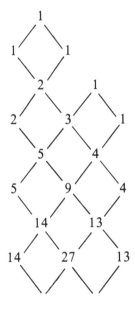

The metaplectic representation of the symplectic groups arises by observing that the character of the Heisenberg projective (irreducible) representation of $(\mathbb{Z}/3\mathbb{Z})^{2m}$ with cocycle given by the usual symplectic form on $(\mathbb{Z}/3\mathbb{Z})^{2m}$ is invariant under the action of the symplectic group; thus one gets a (projective) representation of the symplectic group. It is well known that the representation of $\text{Sp}(2m, \mathbb{Z}/3\mathbb{Z})$ decomposes as a direct sum of two irreducible representations of dimensions $[\frac{3^m}{2}]$ and $[\frac{3^m}{2}] + 1$. We see this pattern on the odd levels of our Bratteli diagrams; so we might suspect that the representation has something to do with the metaplectic representation. In fact the matrix representing a transvection in $\text{Sp}(2n, \mathbb{Z}/3\mathbb{Z})$ may be written in the form $ae_i + b$ where the e_i's satisfy (i), (ii), and (iii) of §3.4.

The group $\text{Sp}(6n, \mathbb{Z}/3\mathbb{Z})$ is essentially the finite simple group of order 25,920, which I discovered by accident when I calculated the matrices for the braid group generators and compared them with those in [**ST**] (see [**J4**]).

There are essentially no more values of m for which the image of the braid group in the e_i algebra is a finite group, except for the curious case $\tau^{-1} = 4\cos^2 \pi/10$ where the representation of B_n is finite for $n = 3$ but not for $n \geq 4$ (the image for $n = 3$ is something like the alternating group A_5).

5.6. Representations in the Pimsner-Popa-Temperley-Lieb algebra (PPTL).
The analysis of §5.5 remained very abstract. Although it is not hard to calculate the matrices for a particular representation if really desired, we gave no general formula. It is possible to give a construction of the representations by paths on the Bratteli diagram ([**Oc2**], [**Sun**], [**Pa**]); this is rarely necessary.

On the other hand, it is satisfying to be able to work with concrete matrices and there is a beautiful representation of the e_i algebra in the continuous series $\tau \leq \frac{1}{4}$, which was discovered first by Temperley and Lieb (see Lecture 8) and rediscovered by Pimsner and Popa [**PP1**] working with subfactors. The idea of the construction is simple and elegant and will recur in greater generality later in these lectures; so we introduce it rather carefully.

First let e_{ij}, $i, j = 1, 2$, be matrix units in $M_2(\mathbb{C})$, i.e., $e_{ij}^* = e_{ji}$, $e_{ij}e_{kl} = \delta_{jk}e_{il}$, $\sum_{i=1}^{2} e_{ii} = 1$. Then we define $e \in M_2(\mathbb{C}) \otimes M_2(\mathbb{C})$ by $e = \frac{1}{1+t}(e_{11} \otimes e_{11} + te_{22} \otimes e_{22} + \sqrt{t}(e_{12} \otimes e_{21} + e_{21} \otimes e_{12}))$. Note that for $\tau \leq \frac{1}{4}$, $t \in \mathbb{R}$; so e is a (rank one) orthogonal projection. Now let $A = M_2(\mathbb{C}) \otimes M_2(\mathbb{C}) \otimes \cdots$ (only finitely many tensor product factors are necessary to represent only finitely many e_i's, but it is convenient to work with the infinite tensor product). The algebra A has an obvious shift endomorphism θ onto $1 \otimes M_2(\mathbb{C}) \otimes M_2(\mathbb{C}) \otimes \cdots$. If $M_2(\mathbb{C}) \otimes M_2(\mathbb{C})$ is identified with the first two tensor product components of A, we may consider e as an element of A. Define $e_i = \theta^{i-1}(e)$, $i = 1, 2, \ldots$. It is clear that $e_i e_j = e_j e_i$ if $|i - j| \geq 2$ and a calculation in $M_2(\mathbb{C}) \otimes M_2(\mathbb{C}) \otimes M_2(\mathbb{C})$ shows with little effort that $e_i e_{i\pm 1} e_i = \tau e_i$.

Thus we can now represent B_n quite concretely on $\bigotimes_{i=1}^{n} M_2(\mathbb{C})$ by sending σ_i to $te_i - (1 - e_i)$.

Some obvious questions arise from the above construction. First one might ask: can it be generalized? We will take this up in §5.7. Another question very relevant to the von Neumann algebras is: is the algebra generated by the above e_i's the same as the one generated by the e_i's in the tower from a pair $N \subseteq M$, $[M : N] = \tau^{-1}$? This question would be settled if we could exhibit on the PPTL e_i's a trace functional tr with property (iv) of §3.4. The (normalized) trace on $\bigotimes M_2(\mathbb{C})$ does *not* have the relevant property. It is a source of constant delight to the author that the relevant linear functional to consider is the Powers state φ_λ of §1.10, so intimately tied up with type III factors! And $t = \lambda$! In fact, it is quite trivial to prove that $\varphi_\lambda(xe_{n+1}) = \tau\varphi_\lambda(x)$ for $x \in \bigotimes_{i=1}^{n+1} M_2(\mathbb{C})$ (with obvious identifications) and also the e_i's belong to the centralizer of the φ_λ; so in particular the restriction of φ_λ to alg$\{1, e_1, e_2, \ldots\}$ is a (positive definite for $\lambda \in \mathbb{R}^+$) trace satisfying (iv) of §3.4. This suffices to prove that the e_i's above generate the same algebra as in a tower since, as we have tried to emphasize, it is the positive definiteness of the trace that decides the structure of the algebra.

The above representation appeared as a "Deus ex Machina" and one might wonder how it could be seen as naturally appearing. To Temperley and Lieb the e_i's occurred as transfer matrices in a certain statistical mechanical model and were, in this sense, natural (see §8.6). This was the direction which generalized in the manner of §5.7. Another way to see the above e_i's arising is by iterating the tower construction for the pair $\mathbb{C} \subseteq M_2(\mathbb{C})$, of finite-dimensional algebras, but using the *Powers state* instead of the trace throughout. The explicit PPTL formulae follow from an appropriate identification of the elements of the tower with the tensor product of 2×2 matrices.*

5.7. QISM representations of the braid group. It is quite natural to look for a straightforward generalization of the PPTL representation simply by replacing "2" by "n" in the construction and looking for an appropriate element $e \in M_n(\mathbb{C}) \otimes M_n(\mathbb{C})$. This was certainly done by Wenzl and Pimsner-Popa. The formula is

$$e = \frac{1}{1+t}\left(t\sum_{i<j} e_{ii} \otimes e_{jj} + \sqrt{t}\sum_{i \neq j} e_{ij} \otimes e_{ji} + \sum_{i>j} e_{ii} \otimes e_{jj}\right),$$

and a calculation not much more difficult than the previous one shows that, with e_i defined in the obviously analogous way, $e_i^2 = e_i$, $e_i e_{i+1} e_i - \tau e_i = e_{i+1} e_i e_{i+1} - \tau e_{i=1}$ and $e_i e_j = e_j e_i$ for $|i - j| \geq 2$. Then sending σ_i to $te_i - (1 - e_i)$ as before gives a representation of the braid group. The structure of the algebra generated by the e_i's in this case will be explained in Lecture 9.

But the way with a future is to regard the e_i's as secondary and simply look for matrices $\check{R} \in M_n(\mathbb{C}) \otimes M_n(\mathbb{C})$ which will give braid group representations by sending σ_i to $\theta^{i-1}(\check{R})$ in $M_n(\mathbb{C}) \otimes M_n(\mathbb{C})$ (where θ is the shift endomorphism as before). It is clear that the only equation that must be satisfied is the one that implies $\sigma_1 \sigma_2 \sigma_1 = \sigma_2 \sigma_1 \sigma_2$. This is conveniently written as

(YBE) $$\check{R}_{12}\check{R}_{23}\check{R}_{12} = \check{R}_{23}\check{R}_{12}\check{R}_{23}$$

where $\check{R}_{12} = \check{R} \otimes \text{id}$ on $(M_2(\mathbb{C}) \otimes M_2(\mathbb{C})) \otimes M_2(\mathbb{C})$ and $\check{R}_{23} = \text{id} \otimes \check{R}$ on $M_2(\mathbb{C}) \otimes (M_2(\mathbb{C}) \otimes M_2(\mathbb{C}))$. This is a watered down version of the Yang-Baxter equation and it has received a great deal of attention. It is also often written as $R_{12}R_{13}R_{23} = R_{23}R_{13}R_{12}$ where R is $P\check{R}$, P being the permutation matrix $\sum e_{ij} \otimes e_{ji}$ in $M_n(\mathbb{C}) \otimes M_n(\mathbb{C}) = \text{End}(\mathbb{C}^n \otimes \mathbb{C}^n)$.

---

*The question of the faithfulness of these representations of the braid groups remains open.

The general solution of (YBE) is far from known. In the $M_2(\mathbb{C})$ case there is the PPTL solution and a few odd solutions and that is it.* In the $M_3(\mathbb{C})$ case, to the best of my knowledge, all solutions are not known though we shall see that there is another one-parameter solution different from the generalized PPTL solution above.

The role of the YBE in statistical mechanics will be discussed in Lecture 8. Suffice it to say that it is of considerable interest to be able to construct as many solutions as possible. The way to do this as part of the quantum inverse scattering method (QISM) is to expand R as a power series in the variable h ($t = e^h$) so that $R = 1 + hr + O(h^2)$. Then YBE implies the equation $[r_{12}, r_{13}] + [r_{13}, r_{23}] + [r_{23}, r_{12}] = 0$. This equation was recognized by Faddeev and his school as occurring in a crucial way in the theory of solitons, and it was known how to construct solutions of it starting from an arbitrary semisimple Lie algebra (note that it makes sense for an arbitrary Lie algebra). The idea was then to "quantize" these "classical Yang-Baxter" solutions. An entirely satisfactory mathematical account of the theory has been worked out by Drinfeld and appears in [**Dr**], though it should be noted that Jimbo obtained many explicit R-matrix formulae [**Ji1**]. I also recommend Faddeev's notes [**F**] for the connection with soliton theory.

The final result is that, to every finite-dimensional representation on V of every simple complex Lie algebra, there is a matrix R in End$(V \otimes V)$ satisfying YBE. The PPTL matrices correspond to sl_2 in its 2-dimensional representation and the generalized ones to sl_n in its n-dimensional identity representation. The ones for general simple Lie algebras are quite complicated and could not be guessed like the generalized PPTL. One needs the formalism of the QISM. In fact, Drinfeld's "quantum double" approach allows him to define a "universal" R matrix for each simple Lie algebra, which can then be specialized to a particular representation.

From the point of view of braid group representations, one may just take the formulae for the R-matrices as a gift from god and try to analyze the ensuing representations. (It would be foolish though to neglect the QISM origins as anyone who has tried even to verify YBE directly will tell you!) The most complete list of explicit formulae that I know of is in [**Res**] where the 49×49 R-matrix for G_2 in its 7-dimensional representation occurs almost explicitly. The R-matrices for sl_2 in all its irreducible representations appear explicitly in [**J5**], these formulae just being a computation of Drinfeld's universal R element of [**Dr**] and Jimbo's explicit formulae for the quantized representations of sl_2 (see [**Ji2**]).

There is a lot to say about these braid group representations and this is an active area right now. I will say more about them in Lecture 8, but for now let me just say that all these representations contain the Burau representation

---

*There is also an interesting solution with parameter arising from the exterior algebra functor applied to the Burau representation.

as a direct summand, but none of them has been obtained in a natural way from the action of the braid group on the free group. We remain at the abysmal "generators and relations" level.

Just to tickle the reader's fancy, here is an explicit R-matrix corresponding to the 3-dimensional representation of sl_2 or $SO(3)$. Note how it is not equivalent to the 3-dimensional generalized PPTL matrix:

$$\begin{bmatrix} 1 & 0 & 0 & 0 & 0 & 0 & 0 & 0 & 0 \\ 0 & 0 & 0 & s^2 & 0 & 0 & 0 & 0 & 0 \\ 0 & 0 & 0 & 0 & 0 & 0 & s^4 & 0 & 0 \\ 0 & s^2 & 0 & 1-s^4 & 0 & 0 & 0 & 0 & 0 \\ 0 & 0 & 0 & 0 & s^2 & 0 & s-s^5 & 0 & 0 \\ 0 & 0 & 0 & 0 & 0 & 0 & 0 & s^2 & 0 \\ 0 & 0 & s^4 & 0 & s-s^5 & 0 & x & 0 & 0 \\ 0 & 0 & 0 & 0 & 0 & s^2 & 0 & 1-s^4 & 0 \\ 0 & 0 & 0 & 0 & 0 & 0 & 0 & 0 & 1 \end{bmatrix}$$

where $x = 1 - s^2 - s^4 + s^6$; here $s^2 = t$ and $i \otimes j$ is the obvious basis of $V \otimes V$. The same matrix occurs in [**AW**] and of course in practically any general paper on YBE solutions derived from QISM.

5.8. The Potts model and Gaussian representations. At the other extreme from the impressive machinery and intricate formulae of the QISM are some simple representations which are, at this stage, entirely computable and can be indulged in by amateurs. But there is every indication that they are quite powerful and their scope is not yet clear. I give a simplified account of work which is joint with D. Goldschmidt and some of which occurs in [**GJ**].

I will base my constructions on the algebra $D_p = \bigcup_n D_{n,p}$ where $D_{n,p}$ is presented on generators $u_1, u_2, \ldots, u_n$ with relations $u_j^p = 1$, $u_j u_{j+1} = \omega u_{j+1} u_j$, $u_j u_k = u_k u_j$ if $|j - k| \geq 2$ (here $\omega = e^{2\pi i/p}$). The sapient reader will have no trouble calculating the Bratteli diagram of D_p:

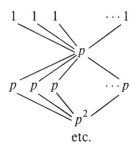

etc.

The idea here is to choose an R-matrix in the algebra generated by u_1. Our D_p also has an obvious shift endomorphism θ and if we ask that $R\theta(R)R = \theta(R)R\theta(R)$, then sending σ_i to $\theta^{i-1}(R)$ gives a representation of the braid group inside D_p, which can easily be made explicit by representing $D_{p,n}$ explicitly.

We know (essentially) of two solutions which coincide when $p = 3$. The *Potts model solution* is $R = (t+1)e - 1$ where $e = \frac{1}{p}(\sum_{k=0}^{p-1} u_1^k)$ and $2 + t + t^{-1} = p$. This solution comes from transfer matrices in the Potts model and is discussed in §8. The other solution is the *Gaussian* $R = \sum_{k=0}^{p-1} \omega^{k^2} u_1^k$. When $p = 4$ there is a solution with a parameter due to the fact that 4 is not prime. When $p = 5$ Goldschmidt has shown that the Potts and Gaussian solutions are essentially the only ones. Much more than that we do not know, although the Gaussian solution can be understood in terms of the metaplectic representation of the symplectic group mod p and generalizes to representations, using the Squier form, inside an algebra with twice as many generators as $D_{n,p}$. See [**GJ**].

In fact, using an appropriate representation, any R element as above can be cast in YBE mould, but this would not be the natural context for it.

5.9. More representations. In fact, many more ways to represent the braid groups are in the air at the moment. In a promising approach, Kohno seems to be able to obtain any YBE solution as the monodromy of a flat connection on the configuration space $(\mathbb{C}^n \setminus \Delta)/S_n$ (See [**Koh1**]). The "interaction round a face" models (e.g. [**ABF**]) give representations closely related to those discussed in §5.7 and §5.8.

But to my mind the most interesting source of representations at the moment is conformal field theory [**BPZ**]. A quantum field theory is defined by its "n-point functions" $\langle \varphi_2(x_1) \cdots \varphi_n(x_n) \rangle$ where φ_i are "fields", x_i are space-time coordinates and the n-point function is supposed to be some kind of correlation density between the various fields at various times. In 2-dimensional conformal field theory the x_i's are pairs of complex numbers z_i and $\bar{z}_i$, but the theory breaks into "chiral" fields $\varphi_i(z)$ depending only on z and one may obtain (I think) the arbitrary n-point functions from those of the form $\langle \varphi_1(z_1), \ldots, \varphi_n(z_n) \rangle$, $z_i \in \mathbb{C}$. The n-point functions have singularities when $z_i = z_j$ for some $i \neq j$, so the above function will only be defined when on $\mathbb{C}^n - \Delta$, whose fundamental group is the pure braid group.

Tsuchiya and Kanie [**TK**] calculated some of these functions for a certain kind of conformal field theory (the Wess-Zumno-Witten model) and found that the chiral n-point functions satisfied differential equations (the Khniznik–Zamolodchikov equations, see [**KZ**]) defining a monodromy representation of the pure braid group. In one case the representations extend to B_n and give precisely the e_i representations at $\tau^{-1} = 4\cos^2 \frac{\pi}{n}$. Moore and Seiberg [**MS**] give a general formalism and seem to be able to replace the braid group by the mapping class group of an arbitrary genus Riemann surface. This raises the interesting prospect of obtaining invariants for 3-manifolds by using the Reidemeister-Singer Theorem [**RS**] in a manner analogous to the way we are about to obtain link invariants from a theorem of Markov.*

---

*This has been achieved in the program laid out by Witten in [**Wi**]. See also [**Koh2**].

Lecture 6. Knots and Links

6.1. Knots and links. By "knot" I will mean a smoothly embedded S^1 in S^3 or $\mathbf{R}^3$. A "link" is a disjoint union of knots. Knots and links may or may not be considered with oriented components. Any link admits a planar projection with singularities of the very simple double-point kind: ✕ . Remembering which part of the curve is closest to the plane of projection at each double point, one has the notion of link diagram. Three examples are drawn below:

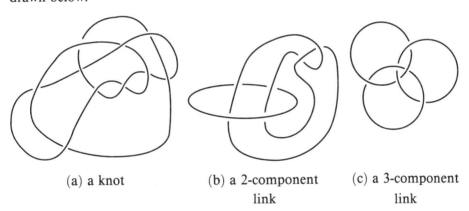

(a) a knot   (b) a 2-component link   (c) a 3-component link

Knots and links are considered up to isotopy in S^3. The translation of isotopy into combinatorics of link diagrams is effected by the Reidemeister moves (and planar isotopy) (see p. 60): the idea being that if ever one of these figures is encountered in a diagram it may be replaced by the one on the other side of the ↔. Reidemeister's Theorem ([**Rei**]) states that two diagrams of isotopic links differ by a sequence of Reidemeister moves. (One needs to consider also the moves obtained from the above by obvious changes of crossings.) If the links are oriented, the moves must include all possible orientations of the strings.

It has recently been emphasized by L. Kauffman that, in fact, the first Reidemeister move is of little significance. Let us call *regular isotopy* the equivalence relation on link diagrams generated by planar isotopy and the types II and III moves. Then a theorem of Trace ([**Tr**]) asserts that two

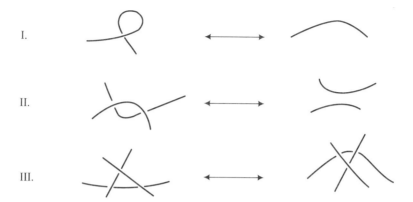

oriented link diagrams that are isotopic are in fact regular isotopic provided they have the same rotation number and writhe (per component) where the *writhe* is the algebraic sum of the crossings, ⤬ being +1 and ⤫ being −1.

As an example of a simple combinatorial invariant, the Reidemeister moves give the linking number of two knots, $\mathrm{lk}(K_1, K_2)$. It may be defined simply as $\frac{1}{2}$ the algebraic sum over all crossings between K_1 and K_2 in some generic projection of the link formed by K_1 and K_2. Thus the linking number for the following diagram is −1:

and for the following is 0:

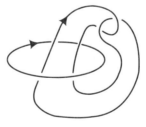

It is a simple exercise to prove that $\mathrm{lk}(K_1, K_2)$ is invariant under the Reidemeister moves. There are plenty of other more intuitive ways to define the linking number, but the above has the advantage of being readily computable from a picture. See [**Rol**] for many equivalent definitions.

6.2. The fundamental group and the Alexander module. The homeomorphism type of the complement of a link is an obvious isotopy invariant of the link. The fundamental group is a powerful tool for studying this complement. Two main presentations are used for $\pi_1(S^3 - K)$ starting from a link diagram.

The Wirtinger presentation takes a point well above the plane as a base point and associates a generator to each overpass, an example of an overpass being as appears with a thick black line below:

The fundamental group element begins at the base point, winds once around its overpass and then returns directly to base, having a linking number of +1 with the knot. One orients the link so that orientation conventions can be established. At every crossing there is a fairly obvious relation, e.g. below, for a positive crossing,

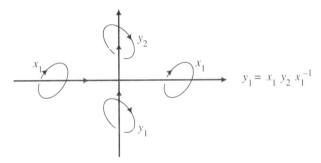

$$y_1 = x_1 y_2 x_1^{-1}$$

and a little van Kampen shows that these generators and relations present $\pi_1(S^3 - L)$.

As an example we suggest that the reader show that the fundamental group of the complement of the trefoil group has presentation $\langle x, y; xyx = yxy \rangle$. Note that this is the 3-string braid group.

Let us now simplify life by talking only about knots for the time being. For the story with links, see any knot theory text, e.g. [**Rol**] or [**BZ**]. The most obvious effect of this simplification is that it means $H_1(S^3 - K) = \mathbb{Z}$ or $\pi_1/[\pi_1, \pi_1] = \mathbb{Z}$, any of the Wirtinger generators being a generator for H_1. Now $\mathbb{Z}$ acts on $[\pi_1, \pi_1] (= \Gamma)$ by conjugation by one of the Wirtinger generators (or any element mapped onto 1 by abelianization). This action on $\Gamma/[\Gamma, \Gamma]$ is independent of which generator is chosen. Thus the abelian group $\Gamma/[\Gamma, \Gamma]$ has a canonical $\mathbb{Z}$-action and becomes a $\mathbb{Z}[\mathbb{Z}] = \mathbb{Z}[t, t^{-1}]$ module, t being conjugation as above. This $\mathbb{Z}[t, t^{-1}]$ module is called the *Alexander module*. More geometrically, it is the first homology of the regular infinite cyclic cover of the complement of the knot (corresponding to the commutator subgroup) as a $\mathbb{Z}[t, t^{-1}]$ module, t being induced by the deck transformation generating the covering group $\mathbb{Z}$. For a link with k (oriented) components, $\pi_1/[\pi_1, \pi_1] = \mathbb{Z}^k$ and the Alexander module becomes a $\mathbb{Z}[t_1, t_1^{-1}, t_2, t_2^{-1}, \ldots, t_k, t_k^{-1}]$ module.

The ring $\mathbb{Z}[t, t^{-1}]$ is quite well behaved (not principal but certainly a UFD), so it is easy to extract information from a module over it. By fiddling with the Wirtinger presentation one may see that the Alexander module of a knot has a *square* presentation matrix Λ. It is not hard to show that the principal ideal of $\mathbb{Z}(t, t^{-1})$ generated by $\det(\Lambda)$ is independent of the presentation matrix Λ. This $\det \Lambda$ is called the *Alexander polynomial* and is, at this stage, only defined up to multiplication by $\pm$ powers of t. It was defined in [**Ax1**].

6.3. Seifert surfaces. From the geometric definition of the Alexander module it is clear that our understanding of it will depend on how good a picture we have of the infinite cyclic cover of the knot complement. The naive way to build pictures of covering spaces is to make cuts in the space and then paste copies together along the cuts so that one copy is a fundamental domain for the group action. The appropriate variety to cut along for a knot complement would be a surface. To be in the right position with respect to the relevant subgroup of the fundamental group the surface should be orientable and its boundary should be the knot. It is also just a plain interesting question as to whether a knot is the boundary of an oriented surface in $\mathbf{R}^3$, and in how many ways.

THEOREM (Seifert). *Every oriented knot or link in $\mathbf{R}^3$ is the (oriented) boundary of an oriented embedded surface in $\mathbf{R}^3$.*

The proof is remarkably simple. Start with an oriented link diagram as below:

Now eliminate every crossing by replacing $\asymp$ by $\asymp$. This has been done above. There remains a finite collection of oriented circles. Each circle bounds an oriented disc and the discs for each circle may be made disjoint by putting all the circles on slightly different levels. Now, in each region where a crossing has been removed, we attach discs with an oriented twisted band as follows:

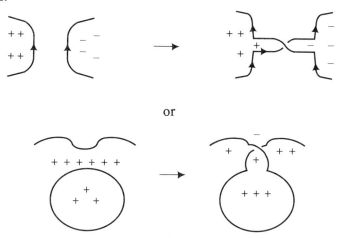

or

This has the effect of reintroducing the crossings, so the boundary of the resulting surface is the link. The whole procedure was oriented so the surface is.

The circles introduced in the above process are called *Seifert circles*.

This settles the existence question. But a given knot may have many inequivalent Seifert surfaces, both as abstract surfaces and as embedded surfaces. Recognizing the abstract surface is easy (though it may be difficult to visualize) because we know that an arbitrary oriented surface with a given number of boundary components is determined by its genus (which is the genus of the surface with the boundary components filled in). This can always be determined by simple homology. The embedding problem is much more difficult.

6.4. Seifert matrices, S-equivalence. Given a Seifert surface M for a link L, its orientation gives a normal nonvanishing vector field on M. Any element of $H_1(M)$ can be represented by a simple closed curve α on M and the *Seifert form* on $H_1(M)$ is given by $(\alpha_1, \alpha_2) = \text{lk}(\alpha_1, \alpha_2^+)$ where α_2^+ is α_2 pushed a little bit off M in the direction of the vector field. Given a basis of $H_1(M)$, the matrix of the Seifert form with respect to this basis is called a Seifert matrix. A Seifert matrix V is obviously only defined up to change of basis $V \mapsto XVX^T$ (X invertible, over $\mathbb{Z}$), but of course different Seifert surfaces may give rise to widely different Seifert matrices. The equivalence relation on matrices V, which says when they are Seifert matrices for the same link, is surprisingly easily described. It is called S-equivalence. It is generated by change of basis as above and a stabilizing move $V \to \begin{pmatrix} V & 0 \\ 0 & \begin{smallmatrix} 0 & 1 \\ 0 & 0 \end{smallmatrix} \end{pmatrix}$.

Thus any invariant of square matrices which is invariant under these two moves gives a link invariant. Perhaps the most obvious is the signature of the matrix $V + V^T$. The resulting knot invariant, called the signature of the knot, is important in connection with four-manifolds.

It is not too hard to see, using a Mayer-Vietoris sequence, that $tV - V^T$ is a presentation matrix for the Alexander module of the link. Thus, in particular, $\det(tV - V^T)$ is the Alexander polynomial of a link with Seifert matrix V.

EXAMPLE. The trefoil.
Seifert surface

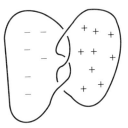

Basis of H_1

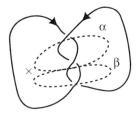

Seifert matrix

$$\begin{array}{c} \alpha \beta \\ \begin{array}{c} \alpha \\ \beta \end{array} \begin{pmatrix} 1 & 1 \\ 0 & 1 \end{pmatrix} \end{array}$$

$$V - tV^T = \begin{pmatrix} t-1 & t \\ -1 & t-1 \end{pmatrix}, \qquad \det = 1 - t + t^2.$$

A more symmetric way to handle this would be to introduce $\sqrt{t}$ and consider $\sqrt{t}V - \frac{1}{\sqrt{t}}V^T$. It is clear that the determinant of this matrix is an invariant of S-equivalence. We see that the Alexander polynomial can be canonically normalized and that $\Delta(t) = \Delta(t^{-1})$ for knots.

One way to make knots with obvious Seifert surfaces is to attach bands to a disc as below.

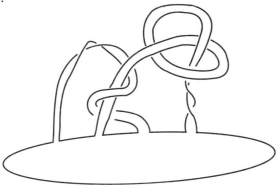

The Seifert matrix can easily be read off such a picture. In this way, Seifert in [Sei] was able to show that the conditions $p(t) = p(t^{-1})$, $p(1) = 1$ on Laurent polynomials, characterize Alexander polynomials of knots.

6.5. Untwisted doubles of knots have trivial Alexander module. There is a very pretty and simple way to construct knots with trivial Alexander module (and hence Alexander polynomial). It is the process of doubling. The idea of doubling a knot is to replace the one string by two and join somewhere by a clasp to be sure of having a knot rather than a link, as below.

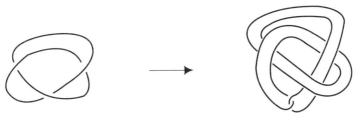

The useful thing is that there is a very obvious Seifert surface which goes around the knot in between the parallel strands and is completed at the clasp as below.

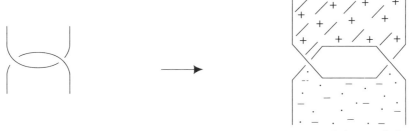

(If this surface is not oriented, just put a twist in the parallel strands.)

This surface clearly has genus one with two obvious homology generators—one, α, which goes just round the clasp, and the other, β, which goes all the way round the band. Note that the unknot may be obtained in this way as below.

On the other hand, the only difference between the Seifert matrix for the doubled knot and that for the above unknot is in the linking of β with its push-off β^+. This can be arranged arbitrarily by giving the band an appropriate number of twists. In particular, we can get the same Seifert matrix as for the unknot. Thus, for instance, the following knot has trivial Alexander module.

6.6. Skein relation for the Alexander polynomial. If L_+, L_-, and L_0 are oriented links having projections identical except near one crossing where they are as below:

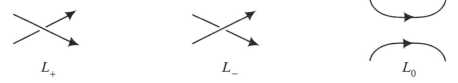

we may look at relations between the Seifert matrices for L_+, L_-, and L_0 for the surface obtained by Seifert's algorithm. Note that because of orientation, the upper and lower regions in the diagram are interior to two distinct Seifert circles. If the Seifert matrix for L_0 is V then that for $L_\pm$ is $\begin{pmatrix} V & A \\ B & x\pm\frac{1}{2} \end{pmatrix}$ where a basis for H_1 of the surface for L_0 has been completed to a basis for $L_\pm$ by some closed curve going across the $L_\pm$ band. Thus for $L_\pm$, the matrix whose determinant is the normalized Δ is

$$\begin{pmatrix} \sqrt{t}V - \frac{1}{\sqrt{t}}V^T & \sqrt{t}A - \frac{1}{\sqrt{t}}B \\ \sqrt{t}B - \frac{1}{\sqrt{t}}A & \left(\sqrt{t} - \frac{1}{\sqrt{t}}\right)(x \pm \frac{1}{2}) \end{pmatrix}.$$

Expanding this determinant by the last row or column, then subtracting, we get

$$\Delta_{L_+} - \Delta_{L_-} = \left(\sqrt{t} - \frac{1}{\sqrt{t}}\right)\Delta_{L_0}.$$

This is known as the skein relation. It is easy to see that this relation suffices to calculate the Alexander polynomials of all links inductively. Although it looks like a slow algorithm for a given link, in practice the skein relation is quite rapid for hand calculations. It was used by Conway in [Co] to calculate Alexander polynomials, and the general consideration of triples L_+, L_-, L_0 as above was introduced under the name of skein theory.

6.7. Closed braids and the Burau representation. There are connections between the theory of braids and that of knots. Roughly speaking, braids are elements of an "algebra" over a "field" which is the knots. There are ways to obtain "field" elements from the "algebra" elements. The first way that we describe here is the "trace" or closure of the braid, which, given $\alpha \in B_n$, we shall denote $\hat{\alpha}$. It is illustrated below.

Note that, up to a convention, the link $\hat{\alpha}$ is canonically oriented. A result of Alexander [Ax2] states that any oriented link in $\mathbf{R}^3$ is the closure of some braid (the "trace" is surjective), although one may need a large number of strings to obtain a given link. We illustrate the procedure with two examples below which really give the general method.

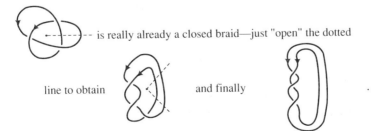

Thus if there is a point in the link diagram for which the link always turns in the same way around the point, you have a closed braid. If there isn't such a point, you just make one, e.g. as below:

Choose the point marked with a • and

now throw the bottom arc over the back:

It is thus tempting to try to use braids and their group structure to study links. The temptation is strengthened by the connection between the procedure for calculating the Alexander module and that for constructing the Burau representation. One may in fact show that, if $\beta(\alpha)$ is the Burau matrix of $\alpha \in B_n$, then $\det(1 - \beta(\alpha)) = (1 + t + \cdots + t^{n-1})\Delta_{\hat{\alpha}}(t)$, see [**Bi1**], where in the case of links the infinite cyclic cover is that coming from the homomorphism $\pi_1(S^3 - L) \to \mathbb{Z}$ given by linking number with the link.

The difficulty with using braids to study links is similar to that encountered using diagrams—a great deal of nonuniqueness in the representation of a link as a closed braid. The theorem that replaces Reidemeister's theorem is due to Markov [**Ma**] (though it is more difficult to prove). For a recent proof, see [**Mor1**]. There are only two "Markov" moves. The first, type I, is algebraically appealing—one replaces the element $\alpha \in B_n$ by $\beta \alpha \beta^{-1} \in B_n$ for any $\beta \in B_n$. It is easy to see that this does not change the closure ("trace" (ab) = "trace" (ba)). The second Markov move is more puzzling—one replaces $\alpha \in B_n$ by $\alpha \sigma_n^{\pm 1} \in B_{n+1}$ (or vice versa). Once again, a diagram suffices to check that this move does not change the closure.

The appealing thing about the first Markov move is that it suggests looking at characters of representations to obtain invariants of links. But of course one would have to find exactly the right representation in order to have invariance under the type II move.

There are other ways to close a braid ("states" rather than "traces"), one of which is called the plat closure, which we shall discuss in the next lecture.

Applying Seifert's algorithm of §6.2 to a braid picture of a knot we see there are as many Seifert circles as there are strings in the braid. Yamada in [**Y**] gives a braiding algorithm for converting an oriented link into a closed braid, which preserves the number of Seifert circles. Thus, two potentially different invariants of knots—minimum number of strings and minimum number of Seifert circles, are in fact the same.

Lecture 7. The Knot Polynomial V_L

7.1. First definition of V_L. As suggested in §6.7, we look for link invariants as characters of representations or at least as linear combinations of such. Let us consider the representations constructed in §5.5 and mix in the trace of §3.4(iv). We see immediately that this equation implies that the trace of a braid behaves in a very simple way under the second Markov move; thus for $\alpha \in B_n$

$$\operatorname{tr}(\alpha\sigma_n) = \operatorname{tr}(\alpha((t+1)e_n - 1)) = (\tau(t+1) - 1)\operatorname{tr}(\alpha)$$
$$= \left(\frac{t}{(1+t)^2}(t+1) - 1\right)\operatorname{tr}(\alpha) = -\frac{1}{1+t}\operatorname{tr}(\alpha),$$

and $\operatorname{tr}(\alpha\sigma_n^{-1}) = -\frac{t}{1+t}$. Thus, if we define

$$V_{\hat{\alpha}}(t) = \left(-\frac{t+1}{\sqrt{t}}\right)^{n-1}(\sqrt{t})^{e(\alpha)}\operatorname{tr}(\alpha)$$

then by Markov's theorem $V_{\hat{\alpha}}(t)$ depends only on $\hat{\alpha}$. Using Alexander's theorem we find that we have constructed an invariant of oriented links, and $V_0(t) \equiv 1$. (Here we are using the same notation for $\alpha \in B_n$ and its image in the e_i algebra and $e(\alpha)$ is the "exponent sum" or image of α in the abelianization of B_n which is $\mathbb{Z}$.)

It is important to note that (i), (ii), (iii), and (iv) of §3.4 suffice to calculate this invariant, just as they sufficed (with positivity of the trace) to determine the set of index values. In fact, this is not an entirely foolish way to calculate, for if we work in the algebra, we need only keep track of the coefficients of the $\frac{1}{n+1}\binom{2n}{n}$ basis elements to calculate α. Thus, if the number of strings is fixed and is not too big (say ≤ 10) we can calculate $V_L(t)$ for closed braids with an enormous number of crossings. In fact, in the light of what is to follow, we should probably think of the above definition of $V_L(t)$ as simply being an explicit formula for this invariant of a closed braid, just as the determinant of 1− Burau matrix gives an explicit formula for the Alexander polynomial of a closed braid.

It is possible to prove a few elementary things about $V_L(t)$ from the above definition. For instance, if L is the k-component unlink, clearly $V_L(t) = (-\frac{t+1}{\sqrt{t}})^{k-1}$. Also, since the PPTL representation at $t = 1$ is really a symmetric group representation we see $V_L(1) = (-1)^{\#\text{components}(L)-1}$. Using the connection with the Shale–Weil representation in §5.5 we deduce that $V_L(e^{i\pi/3})$ is given by $\sum_v \omega^{\langle v,v \rangle}$ where $\omega = e^{2\pi i/3}$, v runs over H_1 (Seifert surface; $\mathbb{Z}/3\mathbb{Z}$), and $\langle , \rangle$ is the symmetrized Seifert form (the sum must be suitably normalized by the genus of the surface); see [GJ], [LL]. Just for fun, here are a couple of sample calculations of V_L by this method:

(i) $L = \text{trefoil} = \sigma_1^3$: $\sigma_1^3 \to ((t^3+1)e - 1)$; so

$$\text{tr}(\sigma_1^3) = \frac{t(t^3+1)}{(1+t)^2} - 1 = \frac{(1-t+t^2)t - 1 - t}{1+t} = \frac{-1 - t^2 + t^3}{1+t};$$

so

$$V_L = \left(\frac{-1+t}{\sqrt{t}}\right)(\sqrt{t})^3 \text{tr}(\sigma_1^3) = t + t^3 - t^4.$$

(ii) $L = \text{figure 8} = \sigma_1 \sigma_2^{-1} \sigma_1 \sigma_2^{-1} \to \{(t+1)e_1 - 1\}\{(t^{-1}+1)e_2 - 1\}\{(t+1)e_1 - 1\}\{(t^{-1}+1)e_2 - 1\} = 16$ terms; Ans $= t^{-2} - t^{-1} + 1 - t + t^2 = V_L(t)$.

The mirror image symmetry is easily seen to correspond to $t \to t^{-1}$ in the invariant. Thus we have shown that the trefoil is not the same as its mirror image. It is also easy to show that the $(t+1)$'s in the denominator cancel and that $V_L(t)$ is a Laurent polynomial in t for a link with an odd number of components and $\sqrt{t}$ times a Laurent polynomial in t for a link with an even number of components. Unlike the Alexander polynomial, it can never be zero. An example of a nontrivial link with the same polynomial as the unlink (of as many components) has not yet been found.

Experimentally V_L is quite useful as a distinguisher of links. The first failure to detect a lack of mirror image symmetry occurs for the knot 9_{42} (numbering as in [Rol]) and the first pair of prime knots not distinguished by V_L are 8_8 and 10_{129}; for the record,

7.2. The theory of plats. If $\alpha \in B_{2n}$ is a braid with an even number of strings we may form the plat closure $\tilde{\alpha}$ as follows.

$\alpha \in B_n$ [diagram] $\longrightarrow \tilde{\alpha} =$ [diagram]

The link $\tilde{\alpha}$ is clearly an unoriented link. To see that every (unoriented) link is a plat is in fact even easier than with braids. One just pushes local maxima to the top and local minima to the bottom of a link diagram. This plat closure corresponds to some kind of "state" on the "algebra" of braids.

There is a theorem due to Birman [**Bi2**] which is analogous to Markov's theorem. It says that if $\alpha \in B_{2n}$ and $\beta \in B_{2n}$ have the same plat closure, then there is a sequence of "Birman moves" which take α to β. The moves are of two kinds. The first is to multiply α on the left or right by an element of the group generated by σ_i (i odd), $\sigma_{2i}\sigma_{2i-1}\sigma_{2i+1}\sigma_{2i}$ ($1 \leq i < n-1$) and $\sigma_{2i}\sigma_{2i-1}^{-1}\sigma_{2i+1}^{-1}\sigma_{2i}$. The second move is a stabilizing move—replace $\alpha \in B_{2n}$ by $\alpha\sigma_{2n} \in B_{2n+1}$, or vice versa. Thus we have another potential approach to finding link invariants. The Birman equivalence classes are unions of double cosets for the subgroups defined above, so one naturally would look for vectors in representations fixed or essentially fixed by this subgroup. This idea, partly due to Dennis Johnson, led to the plat approach to V of §7.3.

The plat picture is intimately related to something topologists call bridges, see [**Rol**]. The minimal number of strings in a plat-closure picture of a link is the minimal bridge number ($\times 2$) which was shown by Milnor to give a lower bound on the total curvature of the link in $\mathbf{R}^3$ ([**Mi**]). It is not a very easy quantity to estimate at this stage.

7.3. A second definition of V_L, the plat approach. The search for vectors fixed by the subgroup of B_{2n} defined by the Birman moves culminated in the rediscovery of the state on the e_i algebra which we are about to describe and which will appear again in §8.

PROPOSITION. *The element $e_1 e_3 \cdots e_{2n-1}$ is a minimal projection in $\mathscr{A}_{2n-1}$.*

Thus we may define a state φ on $\mathscr{A}_{2n-1}$ by $\varphi(x)p = pxp$ with $p = e_1 e_3 \cdots e_{2n-1}$. Note that this state is entirely determined by (i), (ii), and (iii) of §3.4. The state φ is compatible with the inclusions $\mathscr{A}_{2n-1} \subseteq \mathscr{A}_{2n-3}$; so we obtain a state φ on the whole e_i algebra. It is a pure state in the sense that if one performs the G.N.S. construction of §1.5 one obtains an irreducible representation of the e_i algebra.

In the direct sum of a semisimple algebra into simple constituents, a minimal projection can have a nonzero component in only one summand (it "lives" in that summand). It is not hard to see by induction that $e_1 e_3 e_5 \cdots e_{2n-1}$

lives in the extreme left summand of the Bratteli diagram:

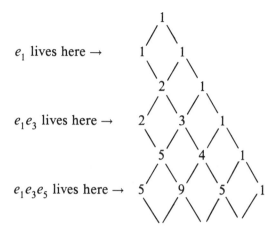

e_1 lives here →

$e_1 e_3$ lives here →

$e_1 e_3 e_5$ lives here →

Moreover, in this simple summand, $e_1 e_3 \cdots e_{2n-1}$ is a rank one projection.

One may show also without difficulty that the pure state φ is related to the trace of §3.4(i) by $\varphi(x) = \frac{1}{\tau^n} \operatorname{tr}(x e_1 e_3 \cdots e_{2m-1})$ for $x \in \mathscr{A}_{2m-1}$.

Identifying where $e_1 e_3$ lives actually makes it quite easy to show that φ is essentially invariant under the Birman moves. To be precise one has that if $\alpha \in B_{2n}$ and x and y belong to the subgroup, $\varphi(x\alpha y) = t^k \varphi(\alpha)$ for some integer $k \in \mathbb{Z}$. Also, φ multiplies by a power of t under the stabilizing move so that we may define an *unoriented* link invariant, defined up to a power of t, as $\varphi(\alpha)$ where $\tilde{\alpha}$ is the link in question. It is shown in [J7] that, in fact, this invariant is just an unnormalized version of $V_L(t)$. The heart of the proof is that a closed n-string braid is much like a $2n$-plat.

A consequence of this formalism is that $V_L(t)$ depends only on the orientation of the link L up to a power of t, which may be identified by known properties of $V'_L(1)$ (see [J6]). There are now a huge number of proofs of this result, which was a surprise when it was first discovered. The plat approach to V_L is closely related to the Kauffman polynomial which we shall introduce later.

7.4. Kauffman's e_i diagrammatics. Pursuing the "algebra", "trace" analogy for knots and braids, Kauffman introduced the following idea: allow the σ_i's in a braid to be extended to braid-like objects E_i where E_i is as below:

In the formalism, a line (no loose ends) is a scalar, so splits off from the picture. Now try working out E_i^2: only 2 strings are relevant, so we get

$$\text{[diagram]} = O\left(\text{[diagram]}\right), \quad \text{i.e.,} \quad E_i^2 = kE_i$$

where k is the scalar corresponding to the unknot. Let us further calculate $E_i E_{i+1} E_i$ and $E_{i+1} E_i E_{i+1}$ (only 3 strings are relevant):

[diagrams]

Thus if we write $e_i = \frac{1}{k} E_i$ we find the same relations as (i), (ii), (iii) of §3.4!!

The explanation of the plat picture is now immediate. For $E_1 E_3 \cdots E_{2n-1}$ is [diagram] and we see that $E_1 E_3 \cdots E_{2n-1} \alpha E_1 E_3 \cdots E_{2n-1}$ is

[diagram with α] = (plat closure of α) × [diagram]

We see that we will be able to calculate V for the plat closures of more general objects than braids as well.

This formalism is quite suggestive.

7.5. Skein relation, third definition of V_L. We have not yet exploited the property that the e_i's are idempotents. This implies immediately that the σ_i's will satisfy a quadratic equation inside the e_i algebra. To be precise $\sigma_i^2 = (t-1)\sigma_i + t$, or alternatively, $(1/\sqrt{t})\sigma_i - \sqrt{t}\sigma_i^{-1} = (\sqrt{t} - \frac{1}{\sqrt{t}})$. This means that if we ever have three braids $\alpha_+ = \beta\sigma_i\beta'$, $\alpha_- = \beta\sigma_i^{-1}\beta'$, and $\alpha_0 = \beta\beta'$, then $\frac{1}{\sqrt{t}}\text{tr}(\alpha_+) - \sqrt{t}\text{tr}(\alpha_-) = (\sqrt{t} - \frac{1}{\sqrt{t}})\text{tr}(\alpha_0)$ so that when we put in the normalization, $t^{-1}V_{\hat{\alpha}_+} - tV_{\hat{\alpha}_-} = (\sqrt{t} - \frac{1}{\sqrt{t}})V_{\hat{\alpha}_0}$. Now it is not hard to convince oneself that a link may be made into a closed braid so that any chosen crossing remains unaltered and turns up as one of the $\sigma_i^{\pm 1}$'s in the braid word. Thus we have proved that if L_+, L_-, and L_0 are links with diagrams identical except near one crossing where they are as below, then $\frac{1}{t}V_{L_+} - tV_{L_-} = (\sqrt{t} - \frac{1}{\sqrt{t}})V_{L_0}$.

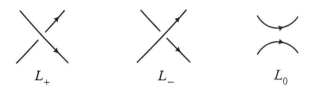

L_+ $\qquad$ L_- $\qquad$ L_0

This forms the basis of a very useful method of computing V_L inductively and proving many facts inductively. For instance, that $\frac{dV}{dt}(1) = 0$ for knots is an easy inductive argument. It is moderately tricky to prove using the e_i algebra definition. Also, $V_K(-1) = \Delta_K(-1)$ is clear from this relation and the one for the Alexander polynomial in §6.6. The polynomial nature of V_L becomes trivial.

Note that the skein relation could also be used as a *definition* of V_L although the existence proof would then be a painful induction.

We invite the reader to calculate the polynomial for the trefoil using the skein relation. The argument that shows that the skein relation suffices to calculate V_L for arbitrary L is the same as that for the Alexander polynomial in §6.6.

7.6. The skein polynomial, inductive definition. Although the skein relation of §7.5 was not terribly appealing as a definition of V_L, juxtaposition of the formulae of §6.6 and §7.5 suggests quite strongly that there may actually be a two-variable generalization of both V_L and Δ_L which embraces them both as specializations. That this is true was proved in [**F+**] and [**PT**] and we will call the resulting three-variable homogeneous Laurent polynomial the *skein polynomial* $P_L(x, y, z)$ defined by $xP_{L_+} + yP_{L_-} + zP_{L_0} = 0$, wherever L_+, L_-, and L_0 are as in §7.5, and $P_0 \equiv 1$. Given the knot tabulations and computer programs, it is not hard to find knots to show that P_L is a more powerful invariant than V and Δ, although it still fails to distinguish 8_8 and 10_{129}.

In some ways P_L is more like Δ than V. It is quite sensitive to changing the orientation of individual components of links (though note that it does not detect a global orientation reversal). A thorough treatment of it from the skein point of view is in [**LM**].

The skein polynomial has been used to do various things. One of them is to give a lower bound for the number of circles produced by Seifert's algorithm (§6.3) for constructing a Seifert surface. The bound is in terms of the degree of one of the variables and was originally due to Morton [**Mo**]. The same bound was found by Morton [**Mo**] and Franks–Williams [**FW**] for the number of strings required for a closed braid form of the link, and Yamada closed the circle by showing that these two invariants, Seifert circle number and braid index, are the same, giving an algorithm for turning a link diagram into a braided link diagram, preserving the number of Seifert circles.

The braid index bound is remarkably sharp for small knots, being exact for all but 5 knots on 10 crossings or less (see [**Jo4**]).

7.7. The Kauffman polynomial. The fact that V_L is essentially an unoriented link invariant suggests looking for invariants of unoriented links. Indeed, Brandt-Lickorish-Millett and Ho proved ([**BLM**], [**Ho**]) that there is an unoriented link polynomial, now known as the *absolute* polynomial $Q_L(x)$, defined inductively by $Q_{\text{unknot}} \equiv 1$ and the generalized skein relation

$$Q_{L_+} + Q_{L_-} = x(Q_{L_0} + Q_{L_\infty})$$

where L_+, L_-, L_0, and L_∞ are links with diagrams differing only near one crossing where they are as below:

L_+ L_- L_0 L_∞.

(Note that this picture is not obviously orientable.)

The axiom for Q provides an algorithm for the calculation of Q by the same reasoning as in §6.6 (L_∞ also has one less crossing than L_+ and L_-).

The absolute polynomial led Kauffman to the discovery of the Kauffman polynomial. The approach is to start with regular isotopy as defined in §6.1 and begin by defining an invariant of regular isotopy for unoriented links, which we call $\widetilde{F}(a, x)$. The axioms for $\widetilde{F}$ are as for Q, $\widetilde{F}_0 \equiv 1$, $\widetilde{F}_{L_+} + \widetilde{F}_{L_-} = x(\widetilde{F}_{L_0} + \widetilde{F}_{L_\infty})$, but since it is not an invariant of isotopy, these axioms no longer suffice to define it. We need to be able to use type I moves to reduce a picture to unknots using the generalized skein theory. But Kauffman postulated a different behaviour under type I moves, namely $\widetilde{F}_{L_\pm} = a^{\pm 1}\widetilde{F}_L$ where $L_+ = \reflectbox{?}$, $L_- = ?$, and L is the link obtained by doing the suggested type I move on L_+ or L_-. One must prove that $\widetilde{F}$ is well defined. This was done by Kauffman in [**K2**].

Once $\widetilde{F}$ is defined one may obtain an invariant for oriented links, the Kauffman polynomial, by renormalizing. Thus if $\vec{L}$ is an oriented link whose unoriented version is L we define

$$F_{\vec{L}}(a, x) = a^{-w(\vec{L})}\widetilde{F}_L(a, x)$$

($w(\vec{L})$ is the writhe as in §6.1).

7.8. Kauffman's "states model", fourth and best definition of V_L. Start with an unoriented link diagram as below:

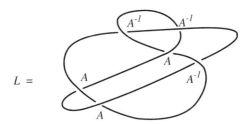

At each crossing one may choose "A" and "A^{-1}" regions according to the following convention: $A^{-1}\diagdown\!\!\!\!\diagup^{\!A}_{\!A}\!\!\diagdown\!\!\!A^{-1}$. A "state" of the link projection is a function from the set of crossings of the projection to the set $\{A, A^{-1}\}$. Given a state σ one "eliminates" all the crossings so as to connect the regions assigned the same symbol by the state; thus $A \to)($, $A^{-1} \to \asymp$. Let $|\sigma|$ be the number of connected components of the resulting picture (e.g., for the diagram and state above one obtains

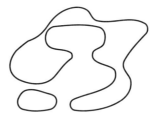

and so $|\sigma| = 2$). Kauffman in [K1] defines $\langle L \rangle \in \mathbb{Z}[A, A^{-1}]$ by $\sum_\sigma A^{A(\sigma)-A^{-1}(\sigma)}(-(A^2+A^{-2}))^{|\sigma|-1}$ where $A(\sigma)$ is the number of "A" crossings defined by σ and similarly $A^{-1}(\sigma)$. It is then simple to check that the bracket $\langle L \rangle$ is unchanged if L is altered by a type II or III Reidemeister move and thus provides an invariant of regular isotopy (see §6.1) of unoriented links. It is also easy to show that if L' and L differ by a type I Reidemeister move then $\langle L' \rangle = (-A)^{\pm 3}\langle L \rangle$ with the sign depending on which kind of type I move is involved. Thus if L is given an orientation we may define an isotopy invariant $\langle\langle \vec{L} \rangle\rangle$ by $\langle\langle \vec{L} \rangle\rangle = (-A)^{-3w(\vec{L})}\langle L \rangle$ where $w(\vec{L})$ is the algebraic sum of the crossings ($\diagup\!\!\!\!\diagdown = +1$, $\diagdown\!\!\!\!\diagup = -1$). One may then show that if L_+, L_-, and L_0 are as in 7.5 then $A^4\langle\langle L_+ \rangle\rangle - A^{-4}\langle\langle L_- \rangle\rangle = (A^{-2} - A^2)\langle\langle L_0 \rangle\rangle$; thus by §7.5, $\langle\langle \vec{L} \rangle\rangle = V_{\vec{L}}(A^{-4})$.

This approach to $V_{\vec{L}}$ has some considerable advantages over the original approach which was outlined in §7.1–§7.7. First of all is the alarming simplicity of the proof. But more important is the very explicit formula for

$V_L(t)$. One may begin to look at the coefficients of the various powers of t and understand where they come from.

For instance, Kauffman and Murasugi have shown that, if one defines the degree of $V_L(t)$ as the difference between the degrees of the monomials of highest and lowest degree occurring in $V_L(t)$, then for any connected link diagram L one has

$$\deg V_L(t) \leq (\text{number of crossings}).$$

Moreover, if L is alternating and not simplifiable in that it looks like the following picture:

$$\boxed{\text{stuff}} \times \boxed{\text{more stuff}},$$

then $\deg V_L(t) = $ (number of crossings). Finally, if L is prime and nonalternating, then $\deg V_L(t) \lneq$ (number of crossings).

For proofs see [Ms], [K3,] [Th], [HKW], [Tu1]. These results have some remarkable consequences concerning alternating knots; in particular, it is clear that a nonsimplifiable (as above) alternating knot cannot be drawn with less crossings than one sees. This was one of the Tait conjectures which had been open for about 100 years. For an account of this see [HKW]. Many other delightful results are proved by Thistlethwaite [Th] who shows that, for an alternating knot, $V_L(t)$ is a specialization of the Tutte-Whitney two-variable dichromatic polynomial of the graph obtained from a chequerboard shading as below:

We shall say more on this in §8.7.

Before leaving the model, note that it follows immediately that $\langle \times \rangle = A \langle \asymp \rangle + A^{-1} \langle\rangle()$. This recursion relation is very useful and really equivalent to the states summation formula which defines $\langle L \rangle$ directly, provided one adds the axiom describing the behaviour of the bracket for disjoint union with the unknot, $\langle L \amalg 0 \rangle = -(A^2 + A^{-2})\langle L \rangle$.

Lecture 8. Knots and Statistical Mechanics

8.1. Statistical mechanics formalism. We will discuss systems made up of a large number of composite subsystems called "atoms". Each of the atoms can be in a certain (small) number of (micro-) states, and a state σ of the whole system is given by the assignment of a microstate to every atom of the system. Each state of the system possesses a certain energy $E(\sigma)$ which is the sum of all energies of interactions between atoms in their given states.

The general principles of equilibrium statistical mechanics say that we should then calculate the so-called "partition function" $Z = \sum_{\text{states } \sigma} \exp(-\beta E(\sigma))$. The idea is that $e^{-\beta E(\sigma)}$ is proportional to the probability of the system occupying the state σ, so that the average value of a physical quantity G will be $\frac{1}{Z} \sum_{\sigma} G(\sigma) \exp(-\beta E(\sigma))$. In order to avoid "transient" phenomena, one would like to consider infinite systems. But of course the sum Z cannot converge for an infinite system. One proceeds by taking a finite approximation P to the whole subsystem with $\#(P)$ atoms. Boundary conditions need to be given to specify P completely. The partition function Z_P then makes sense, and one may form the free energy per atom, $F_P = \frac{1}{\#(P)} \ln Z_P$. Some crude estimates show that F_P ought to have a limit as the finite approximations P tend to the whole system. It is this free energy per site, $F = \lim_{P \to \infty} \frac{1}{\#(P)} \ln Z_P$, that one would like to calculate, given the sequence P and the energy function $E(\sigma)$.

A *model* is a mathematical description of the system S and the energy function $E(\sigma)$. The model is said to be *solved* if the free energy per site F is known explicitly as a function of β and any other parameters, such as a magnetic field. Some simple models in two dimensions have been solved. We give some specific models in the next section.

8.2. Ising, Potts, Vertex, Spin, and IRF models. The most famous of all models is the Ising model (solved in two dimensions by Onsager in [**On**], 1944). We shall restrict ourselves to two-dimensional models. The Ising model consists of atoms placed on the vertices of a two-dimensional lattice.

Each atom can be in two spin states, say +1 or −1. Thus a typical state of the system would be as pictured below:

It is natural enough to use squares or rectangles as the finite approximations of §8.1. Two kinds of boundary conditions are most used: periodic ones and free (i.e., no boundary conditions).

The energy of a state is the sum of the energies from each of the edges, horizontal and vertical. In the state σ, a horizontal edge between vertices x and y contributes $-J\sigma(x)\sigma(y)$ and a vertical edge $-J'\sigma(x)\sigma(y)$. Thus the partition function for a rectangular approximation will be

$$Z = \sum_{\sigma} \exp\left(K \sum_{(x,y)} \sigma(x)\sigma(y) + L \sum_{(x',y')} \sigma(x')\sigma(y')\right)$$

where the first sum is over all horizontal edges, the second sum is over vertical ones and $K = \beta J$, $L = \beta J'$. Writing the exponential as a product, we see that $Z = \sum_{\sigma} \prod_{\text{edges}} w_\varepsilon(\sigma(x), \sigma(y))$, where ε is either "horizontal", H, or "vertical", V, and x and y are the atoms at the two ends of the edge in question, and

$$w_H \text{ is the } 2 \times 2 \text{ matrix } \begin{pmatrix} e^K & e^{-K} \\ e^{-K} & e^K \end{pmatrix},$$

$$w_V \text{ is the } 2 \times 2 \text{ matrix } \begin{pmatrix} e^L & e^{-L} \\ e^{-L} & e^L \end{pmatrix}.$$

Note that the Boltzmann weights $w(\sigma, \sigma')$ depend only on $\sigma + \sigma'$. We could rewrite the partition function as

$$(\text{const}) \sum_{\sigma} \prod_{\text{edges}} w'_\varepsilon(\sigma(x), \sigma(y))$$

where

$$w_H(a, b) = e^{2K\delta(a,b)}, \qquad w_V(a, b) = e^{2L\delta(a,b)}.$$

Forgetting the constant only changes the free energy by a constant.

This formulation is in many ways more satisfactory as it is clear that the labels "+1" and "−1" on the spins are artificial. It also leads immediately to the next model, the so-called Potts model.

The setup for the *Potts model* is exactly as for the Ising model except that there are q "spin" values per site. Thus the model is completely specified by the $q \times q$ matrix of Boltzmann weights. These are, as before,

$$w_H(a, b) = e^{K\delta(a,b)}, \qquad w_V(a, b) = e^{L\delta(a,b)}.$$

It makes sense to generalize the Potts model further to the most general model of this kind which we simply call a *"spin model"*. The setup is exactly as for a q-state Potts model except that the $w(a, b)$ are completely arbitrary.

IRF (interaction round a face) models are somewhat like spin models, having atoms at the vertices of the lattice, but they differ in that the energy of a state is a sum, not over the edges, but over the faces of the lattice. Thus, such a model is defined by giving Boltzmann weights $w(a, b, c, d)$, where a, b, c, d are indices taking on q possible values. The partition function then becomes

$$\sum_\sigma \prod_{\text{faces}} w(\sigma(x_1), \sigma(x_2), \sigma(x_3), \sigma(x_4))$$

where x_1, x_2, x_3, and x_4 are the vertices around a face numbered according to some pre-established convention (Baxter uses $\begin{smallmatrix} x_4 & \square & x_3 \\ x_1 & & x_2 \end{smallmatrix}$ in [**Ba**]).

For *vertex models* the setup is somewhat different. We have a set of q microstates but they are assigned to the edges of the lattice rather than the vertices. Thus a typical state might be

In a state, each atom is surrounded by a configuration like

$$\begin{array}{c} b \\ x \mid y \\ a \end{array}$$

and it responds with an energy whose Boltzmann weight is written $w(a, b \mid x, y)$. The horizontal and vertical directions are already accounted for in the ordering of the indices of w, but it is still possible to allow for different kinds of atoms with different sets of Boltzmann weights.

The partition function then becomes

$$Z = \sum_{\sigma} \prod_{\text{vertices}} w(a, b \mid x, y)$$

where a vertex in the product is surrounded by the configuration as described.

There are many equivalences between the various models, some more interesting than others. We will discuss one in some detail in §8.4. Certainly any model can be made into a vertex model, at the cost of increasing the number of states per site.

8.3. Transfer matrices. The most frequently used techniques for solving models use transfer matrices. This involves inventing a matrix T such that the entries of powers of T are the terms in the partition function. We illustrate first with the one-dimensional Ising model with Boltzmann weights $w(a, b)$, $a, b = 1, 2$. If we consider the 2×2 matrix T whose (a, b)-entry is $w(a, b)$, then the (i, j)-entry of T^n is $\sum w(i_0, i_1) w(i_1, i_2) w(i_2, i_3) \cdots w(i_{n-1}, i_n)$, where the sum runs over all sequences of 1's and 2's with $i_0 = i$, $i_n = j$. Thus the sum of all four entries of T^n gives the partition function of an n-atom one-dimensional Ising model with free boundary conditions. The model is solved simply by finding the largest eigenvalue of T. The trace of T^n gives the partition function with periodic boundary conditions.

In two dimensions we do the same thing except that we make a choice of horizontal and vertical and decide to collapse each horizontal row to an atom which can then be in a large number of states. Thus for the q-state spin model with Boltzmann weights $w_\varepsilon(a, b)$ we consider a rectangular lattice with M rows and N columns. Thus a row, collapsed to a single atom of a linear chain, will have q^N possible states labelled by sequences $(a_1, a_2, \ldots, a_N)$. The transfer matrix T will then be $T_{(a_1, \ldots, a_N), (b_1, \ldots, b_N)} = (\prod_{i=1}^{N-1} w_H(a_i, a_{i+1})) \times (\prod_{i=1}^{N} w_V(a_i, b_i))$ which accounts for all the bonds in the following picture: $\begin{array}{cccc} b_1 & b_2 & & b_N \\ \vert & \vert & & \vert \\ a_1 & a_2 & \cdots & a_N \end{array}$. If we form a q-dimensional vector space $\mathscr{H}$ with basis v_a, we see that T acts on $\bigotimes^N \mathscr{H}$ by

$$T = CD$$

where D is the diagonal matrix

$$D(v_{a_1} \otimes \cdots \otimes v_{a_N}) = \prod_{i=1}^{N-1} w_H(a_i, a_{i+1}) v_{a_1} \otimes \cdots \otimes v_{a_N}$$

and C is the matrix

$$C(v_{a_1} \otimes \cdots \otimes v_{a_N}) = \sum_{b_1, \ldots, b_N} \prod_{i=1}^{N} w_V(a_i, b_i) v_{b_1} \otimes v_{b_2} \otimes \cdots \otimes v_{b_N}.$$

As before the trace of T^M is the partition function Z with periodic vertical boundary conditions, and the sum of all the entries of T^M is Z with free boundary conditions.

For IRF models it seems preferable to use a diagonal-to-diagonal transfer matrix—see [**Ba**], Chapter 13.

For vertex models the transfer matrix picture is slightly different. One thing is the need to account for both horizontal and vertical boundary conditions. Given the Boltzmann weights $w(a, b \mid x, y)$ with a, b in some set X_H and x, y in some set X_V we will form vector spaces $\mathscr{H}$ and $\mathscr{K}$ with bases $\{v_a \mid a \in X_H\}$ and $\{w_x \mid x \in X_V\}$ respectively. We then define the matrix $\tilde{T}$ on $(\bigotimes^N \mathscr{H}) \otimes \mathscr{K}$ by

$$(*)\quad \tilde{T}(v_{a_1} \otimes \cdots \otimes v_{a_N} \otimes w_x)$$
$$= \sum_{\substack{b_1,\ldots,b_N,\\ x_1,\ldots,x_{N-1},y}} w(a_1, b_1 \mid x, x_1) w(a_2, b_2 \mid x_1, x_2) \cdots w(a_N, b_N \mid x_{N-1}, y) v_{b_1}$$
$$\otimes v_{b_2} \otimes \cdots \otimes v_{b_N} \otimes w_y.$$

$\tilde{T}$ may be thought of as a $\dim(K) \times \dim(K)$ matrix of elements of $\operatorname{End}(\bigotimes^N \mathscr{H})$. It is called the *monodromy matrix*. The sum of its diagonal entries, T, is easily checked to be the row-to-row transfer matrix for the vertex model with periodic horizontal boundary conditions. The usefulness of seeing it as the trace of a monodromy matrix will be seen in §8.5.

8.4. The six-vertex model, Temperley–Lieb equivalence. The six-vertex model is a special vertex model with two states per edge, solved by Lieb in [**Li**]. Temperley–Lieb equivalence ([**TL**]) is an equivalence between the six-vertex model and the Potts models for all values of q, but taken at a special value (criticality) of their parameters. We shall give the proof of [**TL**] as adapted in [**Ba**], Chapter 12.

Unfortunately, the transfer matrices will appear slightly different from those described in §8.3 for vertex models—we use a diagonal-to-diagonal way of generating the lattice.

For the moment we keep the general vertex model setup of §8.3. We approach the lattice as below; we allow ourselves to be a bit sloppy about

$2N$, $2N-1$, etc.,

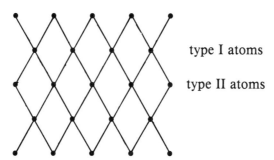

type I atoms

type II atoms

and want to write a transfer matrix to go from

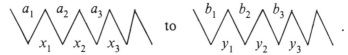

It is clear that we should form a $q^{\dim \mathscr{H} + \dim \mathscr{H}}$-dimensional vector space $\mathscr{H} \otimes \mathscr{H}$ with basis $v_{a,x}$ and define a transfer matrix for a row (as in the above figure—a diagonal from the point of view of §8.3) of type I atoms by

$$T_1(v_{a_1,x_1} \otimes \cdots \otimes v_{a_N,x_N}) = \sum_{b_i,y_i} \prod w_I(a_i, b_i \mid x_i, y_i) v_{y_1,b_1} \otimes \cdots \otimes v_{y_N,b_N}.$$

Observe that this T_1 is just a product $\prod_{i=1}^{N} R_{2i}$ where $R \in \operatorname{End} \mathscr{L}$ is given by $Rv_{a,x} = \sum w_I(a, b \mid x, y) v_{b,y}$, and we have used the notation of the Appendix. To go from the type II atoms back to the type I atoms one uses $T_2 = \prod_{i=1}^{N} R_{2i-1}$. Thus if we do not worry too much about boundary conditions the relevant transfer matrix is

$$(R_1 R_3 R_5 \cdots R_{2N-1})(R_2 R_4 \cdots R_{2N}).$$

Let us now return to the Potts model where $T = CD$, $C(v_{a_1} \otimes \cdots \otimes v_{a_N}) = \sum_{b_1,\ldots,b_N} \prod_{i=1}^{N} w_V(a_i, b_i) v_{b_1} \otimes \cdots \otimes v_{b_N}$, and D is diagonal. Then we notice that both C and D factor into products of commuting matrices conveniently expressed as follows:

for D, define $U_{2i}(v_{a_1} \otimes \cdots \otimes v_{a_N}) = \sqrt{q}\, \delta(a_i, a_{i+1})(v_{a_1} \otimes \cdots \otimes v_{a_N})$,

for C, define $U_{2i-1}(v_{a_1} \otimes \cdots \otimes v_{a_N}) = \dfrac{1}{\sqrt{q}} \sum_b v_{a_1} \otimes \cdots \otimes v_b \otimes \cdots \otimes v_{a_N}$

(this particular form of U_i is chosen because of the special Boltzmann weights for the Potts model). Then we find immediately that

$$C = \prod_{j=1}^{N}((e^K - 1) + \sqrt{q}\, U_{2i-1})$$

and
$$D = \prod_{j=1}^{N-1}\left(1 + \frac{(e^L-1)}{\sqrt{q}} U_{2i}\right).$$
Thus if we define
$$R_{2i-1} = (e^K - 1) + \sqrt{q}\, U_{2i-1}$$
and
$$R_{2i} = 1 + \frac{e^L - 1}{\sqrt{q}} U_{2i}$$
we find that the row-to-row transfer matrix with free boundary conditions is
$$(R_1 R_3 \cdots R_{2N-1})(R_2 R_4 \cdots R_{2N}).$$
Now the U_i's satisfy the following relations:
$$U_i^2 = \sqrt{q}\, U_i, \qquad U_i U_{i\pm 1} U_i = U_i, \qquad U_i U_j = U_j U_i \ \text{if}\ |i-j| \geq 2.$$

Thus, after renormalization $e_i = (1/\sqrt{q}) U_i$, we get the same relations as in §3.4. We can use the second lemma of §3.5 to conclude that the product $U_1 U_3 \cdots U_{2N-1}$ defines a linear functional φ on the algebra generated by the U_i's. But for the Potts model, it is easily seen that $U_1 U_3 \cdots U_{2N-1}$ is a matrix, all of whose entries are a power of $1/\sqrt{q}$. Thus the functional φ evaluated on T is, up to an easily calculated power of $\sqrt{q}$, the partition function for the system with free boundary conditions. We see that the partition functions for the six-vertex model (with some boundary conditions) and the self-dual Potts model are thus equivalent to one another under a simple change of variables, where self-duality means that $(e^L - 1)(e^K - 1) = q$.

Baxter et al have given a beautiful direct argument which shows that Temperley–Lieb equivalence works on an arbitrary planar graph. See [Ba], Chapter 12. As an intermediary in the proof, one finds exactly the Kauffman states model of §7.8 for V_L, defined on a link projection!

8.5. Commuting transfer matrices, the Yang–Baxter equation. Let us suppose that we have a square lattice model and we want to find an explicit expression for the free energy per site in the infinite limit. Then, as we saw in the one-dimensional case, we should try to diagonalize the row-to-row transfer matrix T. Unfortunately, the size of this matrix is tending to infinity, so direct diagonalization is impossible. If, on the other hand, we knew a large family of matrices commuting with T, any eigenspace of T would have to be invariant under this family also. Thus there would be many constraints on the eigenvectors of T, which should help find them. Baxter's idea is to use theories in which T depends on a parameter (known as the "spectral parameter") λ, and the $T(\lambda)$ commute with each other for different values of λ. Is there a simple equation on the Boltzmann weights $w(a, b \mid x, y)(\lambda)$ which guarantees this?

To answer this we recall the formalism of §8.3, for vertex models, where in order to have freedom with the boundary conditions we defined a monodromy matrix $\tilde{T}(\lambda)$ which is a $\dim(\mathcal{H}) \times \dim(\mathcal{H})$ matrix of endomorphisms of $\bigotimes^N \mathcal{H}$, $\mathcal{H}$ (resp. $\mathcal{K}$) being a vector space with basis indexed by the horizontal (resp. vertical) spins per edge. The sum of the diagonal entries of $\tilde{T}(\lambda)$ is $T(\lambda)$. Now suppose that there exists a family $R(\lambda, \mu)$ in $\mathrm{End}(\mathcal{K} \otimes \mathcal{K})$ with the property

$$(*) \qquad R^{-1}(\lambda, \mu)\tilde{T}_1(\lambda)\tilde{T}_2(\mu)R(\lambda, \mu) = \tilde{T}_2(\mu)\tilde{T}_1(\lambda)$$

where we have used the notation of the appendix. Now let E be the "trace" from $\mathrm{End}(\bigotimes^N \mathcal{H}) \otimes \mathrm{End}(\mathcal{K} \otimes \mathcal{K})$ to $\mathrm{End}(\bigotimes^N \mathcal{H})$. Then $E(\tilde{T}_1(\lambda)) = E(\tilde{T}_2(\lambda)) = T(\lambda)$ and a simple calculation shows that equation $(*)$ implies, by applying E to both sides,

$$T(\lambda)T(\mu) = T(\mu)T(\lambda).$$

Applying $(*)$ several times to $\tilde{T}_1(\lambda)\tilde{T}_2(\mu)\tilde{T}_3(\nu)$ we find that

$$[R_{12}(\lambda, \mu)R_{13}(\lambda, \nu)R_{23}(\mu, \nu)][R_{23}(\mu, \nu)R_{13}(\lambda, \nu)R_{12}(\lambda, \mu)]^{-1}$$

is a $(\dim \mathcal{K})^3 \times (\dim \mathcal{K})^3$ matrix of scalars that commutes with $\tilde{T}_1(\lambda)\tilde{T}_2(\mu)\tilde{T}_3(\nu)$. The simplest possibility is then that this matrix is the identity:

(YBE1) $\qquad R_{12}(\lambda, \mu)R_{13}(\lambda, \nu)R_{23}(\mu, \nu) = R_{23}(\mu, \nu)R_{13}(\lambda, \nu)R_{12}(\lambda, \mu).$

If $R(\lambda, \mu)$ is actually a function of $\lambda - \mu$, written $R(\lambda - \mu)$, we obtain

(YBE2) $\qquad R_{12}(\lambda)R_{13}(\lambda + \mu)R_{23}(\mu) = R_{23}(\mu)R_{13}(\lambda + \mu)R_{12}(\lambda).$

The advantage of this version is that R itself can be used as T, using $\mathcal{K}$ for $\mathcal{H}$, for then in $(*)$ $\tilde{T}_1(\lambda) = R_{12}(\lambda)$, $\tilde{T}_2(\mu) = R_{13}(\mu)$, and $R(\lambda, \mu) = R_{23}(\lambda - \mu)$. This is the original situation used in [**Ba**]. In fact, Baxter assumed $\mathcal{H} = \mathcal{K}$ and that $R(\lambda, \lambda')$ was actually $T(\lambda'')$ and found that he could parametrize things (using elliptic functions) so that $\lambda'' = \lambda - \lambda'$.

At this stage many solutions to $(*)$ and (YBE2) are known. A complete classification is not yet in sight. Note that there are two differences between (YBE2) and (YBE) of §5.7. The first is the arrangement of the subscripts, but this is explained in §5.7. The second is the absence of the spectral parameter in §5.7. We see that if we want to obtain braid group representations from a solution of (YBE2) we must either let $\lambda = \mu = 0$ or $\lambda = \mu = \infty$. The first case invariably gives a triviality, but the second case often gives us the representations we are interested in, provided we let $\lambda \to \infty$ appropriately. See [**ADW**].

8.6. Vertex models on link diagrams. So far in this chapter we have focused on statistical mechanics. Connections with knot theory have been suggested twice. First of all, we saw that the proof of Temperley–Lieb equivalence invoked the algebra we originally used to define the knot polynomial $V_L(t)$. Second, the search for commuting transfer matrices led us to the Yang–Baxter equation, which resembles the braid group presentation. But there are quite direct ways to use the statistical mechanical formalism in knot theory as we now describe.

The idea is quite simple and was first devised in [**J8**]. One simply uses a knot diagram in place of the square lattice in the definition of a vertex model. So for the figure 8 knot a typical state of a system with 3 states per edge would be

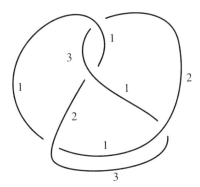

There are clearly 3^8 states in all. We give ourselves two sets $w_\pm(a, b \mid x, y)$ of Boltzmann weights defining the system and agree to use w_+ at a positive crossing ($\times$) and w_- at a negative one ($\times$). The partition function will then be $Z_L = \sum\limits_{\text{states } \sigma} \prod\limits_{\text{crossings}} w_\pm(a, b \mid x, y)$ where for a given crossing the state σ defines the configuration a, b, x, y as follows:

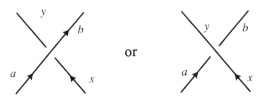

If L had several components we could use different Boltzmann weights at each kind of crossing between the various components. In this way it is possible to treat the different components of a link as having labels.

Of course, for an arbitrary choice of Boltzmann weights we would not expect Z_L to have any interest for the knot theorist. However, one can ask the question: are there conditions on the $w_{\pm}$ such that Z_L as above is a link invariant, i.e., does not change under the Reidemeister moves of §6.1. Indeed, if we play the "regular isotopy" game (§6.1) we see that we really only need to look at type II and III moves. Before doing this we introduce an extra ingredient which will be necessary to get rapidly to the main invariants, and uses, albeit rather weakly, the "spectral parameter" version of the Yang-Baxter equation. A slightly different approach and formulas for many R-matrices appears in [**Tu2**].

Given a state σ we alter the contribution of that state to the partition function by multiplying it by the factor $e^{\hbar \int_L \sigma d\theta}$ where σ is regarded as a locally constant function on L and $d\theta$ is the "change of angle" 1-form on L. Thus the partition function is now

$$Z_L = \sum_{\text{states}} \left(\prod_{\text{crossings}} w_{\pm}(a, b \mid x, y)(\lambda) \right) e^{\hbar \int_L \sigma d\theta}$$

where λ is the angle made by the incoming curves at the crossing:

One may now ask: under what conditions on $w_{\pm}(\lambda)$ is Z_L actually an isotopy invariant of oriented links?

The conditions are straightforward but a little complicated. One simply writes down the conditions necessary for Z_L to be invariant under planar isotopy and each of the Reidemeister moves (for details see [**JSM**]). The most important condition comes from the type III move and is *precisely* (YBE2) of §8.5 for the family $R(\lambda) \in \text{End}(\mathcal{H} \otimes \mathcal{H})$ defined by

$$R(\lambda)(v_a \otimes v_x) = \sum_{b,y} w_+(a, b \mid x, y)(\lambda) v_b \otimes v_y,$$

$\mathcal{H}$ being a vector space with basis $\{v_a\}$.

These conditions would not be too interesting if there were no solutions. But in fact there are many. The one appearing in [**J5**] is

$$w_+(a, b \mid x, y)(\lambda) = \begin{cases} 0 & \text{if } \{a, x\} \neq \{y, b\} \text{ or } a < b, \\ e^{h/2} & \text{if } a = b = x = y, \\ 1 & \text{if } a = b, \ x = y, \ a \neq x, \\ 2\sinh(\frac{h}{2}) \exp(\lambda \, \text{tr}(b-a)) & \text{if } a > b, \ a = y, \ b = x, \end{cases}$$

where a, b, x, y run from $-n$ to $+n$ in steps of 2, and $w_-(a, b \mid x, y)(\lambda, h) = w_+(a, b \mid x, y)(\lambda, -h)$.

This solution was developed from braid group representations generalizing the PPTL representation of §5.6. For $n = 1$ one obtains $V_L(e^h)$ and for general n it is easy to show that the partition function invariant from the above model satisfies

$$e^{\frac{(n+1)h}{2}} Z_{L_+} - e^{-\frac{(n+1)h}{2}} Z_{L_-} = \sinh\left(\frac{h}{2}\right) Z_{L_0}.$$

So we obtain an infinite sequence of specializations of the skein polynomial of §7.6, sufficient to determine it.

To obtain many more solutions of the equations and hence link invariants one may use the machinery of quantum groups which we began to describe in §5.7 ([**Dr**], [**Ji**], [**Wo**]). In Drinfeld's approach one defines a "quasi-triangular Hopf algebra" to be a Hopf algebra A with comultiplication Δ and antipode such that there is an invertible element $R \in A \otimes A$ such that

1) $\sigma(\Delta(x)) = R\Delta(x)R^{-1}$ ($\sigma \in \text{Aut}(A \otimes A)$ is $\sigma(x \otimes y) = y \otimes x$).
2) The natural map $A \to A^*$ defined by R is an algebra homomorphism and coalgebra antihomomorphism.

These two conditions imply a "universal Yang–Baxter equation" $R_{12}R_{13}R_{23} = R_{23}R_{13}R_{12}$, valid in $A \otimes A \otimes A$. To obtain $R(\lambda)$ one may change R by an appropriate one-parameter automorphism group of $A \otimes A$.

Drinfeld also shows how to obtain examples of such quasitriangular Hopf algebras as deformations of the enveloping algebras of simple (finite-dimensional) Lie algebras, using what he calls the "quantum double" construction to obtain the existence of the R element.

An advantage of the universal R matrix approach is that one may treat the components of the link as being labelled. The point is that one may "colour" the link components with finite-dimensional representations of the Hopf algebra. Then at a crossing between a component labelled ρ and another labelled μ one may use as Boltzmann weights the entries of the matrix $\rho \otimes \mu(R(\lambda))$.

Of course one must also worry about type II Reidemeister moves, but Rosso has shown in [**Ros1**] that the universal formulation can be used to handle this as well. There remains only the construction of finite-dimensional representations of the algebra A. If A is a deformation of a simple Lie algebra one can expect the representation to deform also. That this is the case is shown in [**Ros2**], [**Lu**].

Thus we emerge with the following picture: to each simple Lie algebra (over $\mathbb{C}$) there is an invariant of coloured links, which assigns to each labelling of the components with representations of $\mathscr{G}$ a function of the deformation parameter.

8.7. Spin models on link diagrams. Instead of defining vertex models on link diagrams, one may use models generalizing the Potts model. Taking the cue from Temperley–Lieb equivalence and §7.8, one uses the chequerboard shading of the link diagram as the graph for the model. Thus, in the diagram below, the Conway knot is converted to a planar graph with signed edges, the signs coming from the shading thus:

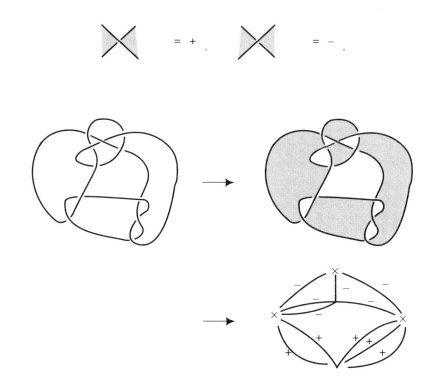

On the ensuing graph one defines a system with q "spin" states per vertex and two sets of Boltzmann weights $w_\pm(a,b)$ which are used to define the partition function

$$Z_L = \sum_{\text{states}} \prod_{\text{edges}} w_\pm(a,b)$$

where "a" and "b" are the spin states of the vertices at the ends of an edge in the product. Everything is delightfully simpler than in the vertex case, so we write down explicitly the equations on $w_\pm$ that guarantee invariance under the type II and III Reidemeister moves:

(S1) $w_\pm(a,b) = w_\pm(b,a) = w_\mp(a,b)^{-1}$ ($w_\pm(a,b)$ is never zero),

(S2) $\sum_b w_+(a,b) w_-(b,c) = q\delta(a,c)$,

(S3) $\sum_x w_+(a,x) w_+(b,x) w_-(c,x) = \sqrt{q}\, w_+(a,b) w_-(b,c) w_-(c,a)$.

For small values of q, say ≤ 8, it seems entirely plausible that one can know all solutions of these equations.

The first solution (for all q) of any real interest is the "Potts model" where

$$w_+(a,b) = \begin{cases} 1 & \text{if } a = b, \\ -t^{-1} & \text{if } a \neq b, \end{cases} \quad 2 + t + t^{-1} = q.$$

The resulting invariant (of unoriented links) is a slightly disguised version of V_L, as can be shown by skein theory.

If the set of all spin states has some structure, one may impose conditions on the Boltzmann weights and look for solutions that satisfy them. In the case where the states are the cyclic group $\mathbb{Z}/q\mathbb{Z}$ there is a solution discovered in [GJ] and [J5]: $w_+(a, b) = Ke^{\pi i(a-b)^2/q}$ where $K^{-2} = \sum_{a \in \mathbb{Z}/q\mathbb{Z}} e^{2\pi i a^2/q}$. The invariant defined by the partition function in this case can be identified in terms of classical (i.e., pre 1984) knot invariants, as done in [GJ]. It is essentially the order of the 1-homology of the 2-fold cover of S^3, branched over the link, with coefficients $\mod q$.*

---

* F. Jaeger has recently discovered a spin model where the natural symmetry group of the Boltzmann weights is the Higman-Sims simple group of order 44,352,000. The model gives a special value of the Kauffman polynomial.

Lecture 9. The Algebraic Approach

9.1. The Hecke algebra. One of the most powerful ways of handling the knot polynomials and constructing subfactors has been to consider and, if necessary, invent algebras generalizing and containing the e_i algebra introduced in Lecture 2. We shall deal first with the Hecke algebra.

The term "Hecke algebra" as used today is very general, being some kind of replacement for the quotient group by a nonnormal subgroup. Given groups $H < G$ the associated Hecke algebra is by definition the commutant of the representation of G acting naturally on the space of functions from G/H to $\mathbb{C}$ or the representation of G obtained by inducing the trivial representation of H up to G.

If we let G be $\mathrm{GL}(n, q)$ (q a prime power) and $H (= B)$ be the subgroup of upper triangular matrices, it follows from abstract nonsense that the Hecke algebra is the linear space of convolution operators by characteristic functions of double cosets. Elementary linear algebra shows that the double coset space in this example is indexed by the symmetric group S_n; indeed, we have $\mathrm{GL}(n, q) = \coprod_{w \in S_n} BwB$ where the symmetric group sits in $\mathrm{GL}(n, q)$ in the obvious way. (This is a simple example of the "Bruhat decomposition".) A natural set of generators for the symmetric group are the transpositions s_i, $i = 1, \ldots, n-1$, between adjacent basis elements. There is a suitable normalization g_i of the characteristic function of Bs_iB so that the following three relations hold:

$$\text{(H1)} \quad g_i^2 = (q-1)g_i + q,$$
$$\text{(H2)} \quad g_i g_{i+1} g_i = g_{i+1} g_i g_{i+1},$$
$$\text{(H3)} \quad g_i g_j = g_j g_i \quad \text{if } |i - j| \geq 2.$$

(For details see [**Bo**].)

One may now let q be an arbitrary complex number ($\neq 0$) and consider the algebra having (H1), (H2), and (H3) as its defining relations. From now on this is what we shall mean by the term "Hecke" algebra, and we shall denote it $H(q, n)$. Relation (H1) means that in any representation of the Hecke algebra the g_i's (all conjugate by (H2)) can only have eigenvalues q

and -1. Thus an alternative way to think of our Hecke algebra is as the classifying object for representations of the braid group such that the usual generators have at most two eigenvalues (which can easily be normalized to be q and -1 for some q).

The most obvious way to treat $H(q, n)$ is as a deformation of the group algebra of the symmetric group S_n. Indeed, for $q = 1$ it is obviously this algebra since the relations (H1), (H2), (H3) then become a presentation of the symmetric group. Now an arbitrary element of the symmetric group can be written as a word of minimal length on the generators $s_1, \ldots, s_{n-1}$ and one may list $n!$ such minimal words which give all the elements of S_n. But starting from a word on the g_i's one may use the same procedure as for the symmetric group to reduce it to a word of minimal length. Use of relation (H1) will reduce the length of a word just as well as $s_i^2 = 1$, the only price paid being that we will get a linear combination of words of minimal length rather than a single one. Thus $\dim(H(q, n)) \leq n!$ for any n. Moreover, from our original definition of $H(q, n)$ for q a prime power we know that $\dim(H(q, n)) = n!$ for these values of q, and hence by polynomial arguments for all values of q. A refinement of these arguments shows that $H(q, n) \cong \mathbb{C}S_n$ for all values of q for which $H(q, n)$ is semisimple. For details see [**Bo**]. A relative version of the argument shows that for all such values of q, the inclusion $H(q, n) \subseteq H(q, n+1)$ will have the same Bratteli diagram as $\mathbb{C}S_n \subseteq \mathbb{C}S_{n+1}$. Using the well-known theory of representations of S_n we obtain the following Bratteli diagram for $H(q, \infty) = \bigcup_n H(q, n)$, for "generic" q (certainly q a prime power). (We have indexed the representations by Young diagrams or partitions which should make the pattern clear.) (See Figure 9.1.)

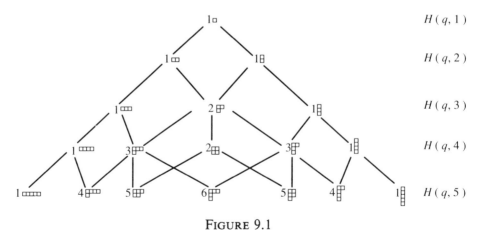

FIGURE 9.1

9.2. The relationship between the e_i algebra and $H(q, n)$. We saw in §5.5 that if $e_1, e_2, \ldots$ satisfy $e_i^2 = e_i$, and $\tau^{-1} = 2 + q + q^{-1}$ then $e_1 e_2 e_1 = \tau e_1$

and $e_2e_1e_2 = \tau e_2$ imply $g_1g_2g_1 = g_2g_1g_2$ for $g_i = (q+1)e_i - 1$. But it is also true that $g_i^2 = q^2 e_i + (1-e_i) = (q-1)((q+1)e_i) + q = (q-1)g_i + q$. Thus the e_i algebra is a homomorphic image of the Hecke algebra. On the other hand, if $e_i^2 = e_i$, $\tau^{-1} = 2 + q + q^{-1}$, and $g_i = (g+1)e_i - 1$, then $g_1g_2g_1 = g_2g_1g_2$ only implies $e_1e_2e_1 - \tau e_1 = e_2e_1e_2 - \tau e_2$ so that the kernel of the map from the Hecke algebra onto the e_i algebra will be the 2-sided ideal generated by $e_1e_2e_1 - \tau e_1$. We can identify this ideal precisely by looking at the Bratteli diagrams, and comparing that for the Hecke algebra and the one of §4.2. We see that the ideal will be that defined by all Young diagrams having three rows or more (or columns depending on convention).

There are a number of ways to make the last deduction rigorous. One way is to analyze the situation at $q = 1$ using the representation theory of the symmetric group and use a deformation argument for other values of q for which the algebras are semisimple. For a complete argument see [GHJ].

9.3. Ocneanu's trace on $H(q, n)$. Thinking of braids let us rewrite $g_i^2 = (q-1)g_i + q$ as $g_i - qg_i^{-1} = (q-1)$ and represent this diagrammatically

$$\diagup\!\!\!\!\diagdown - q \diagdown\!\!\!\!\diagup = (q-1) \asymp.$$

Thus if we were looking for a two-variable extension of V_L we would be naturally led to try to define a trace on the Hecke algebra with another parameter independent of q. We can use Markov's theorem as a guide and ask for the existence of a trace tr: $H(q, \infty) \to \mathbb{C}$ with

1) $\operatorname{tr}(1) = 1$,
2) $\operatorname{tr}(ab) = \operatorname{tr}(ba)$,
3) $\operatorname{tr}(xg_{n+1}) = z\operatorname{tr}(x)$ for x in $H(q, n+1)$.

This is actually how Ocneanu discovered the skein polynomial in [F+].

In fact it is not at all difficult to prove by induction the existence of such a trace. The only important ingredient is that $\dim H(q, n) = n!$, which allows us to define tr freely on basis elements using 1), 2), and 3). For details see [J7].

We deduce immediately the existence of an invariant $X_L(q, \lambda)$ defined by

(X) $$X_L(q, \lambda) = (-(1-\lambda q)/\sqrt{\lambda}(1-q))^{n-1} \operatorname{tr}(\pi(\alpha))$$

where π is the obvious representation of B_n into $H(q, n)$ and $\alpha \in B_n$ is a braid with closure $\hat{\alpha}$, and we have used the variable $\lambda = (1-q+z)/qz$. The relation (H1) shows that X_L satisfies the skein relation:

$$(\lambda q)^{-1/2} X_{L_+} - (\lambda q)^{1/2} X_{L_-} = (\sqrt{q} - 1/\sqrt{q}) X_{L_0},$$

where L_+, L_-, and L_0 are as in §7.5. (One needs to know also that it is possible to make a link into a braid leaving intact anything that already looks like a braid, in particular a single crossing.)

Simply by examining the large q behaviour of definition (X) one may deduce the powerful inequality of Morton-Franks-Williams ([**FW**], [**Mor2**]) which says that, given a link L, the smallest number of strings required to represent L as a closed braid is at least as big as $d_+ - d_- + 1$, where d_+ is the degree of the largest power of λ in X_L and d_- is the smallest.

This gives a lower bound on the number of Seifert circles by Yamada's result (see §6.7).

Another interesting use of the Hecke algebra approach is as a calculational device. It is clear that if one restricts oneself to closures of braids in B_n one may find the element of the Hecke algebra representing the braid simply by recording the changes in the coefficients of $n!$ basis elements as one multiplies by the Hecke algebra generators for a word defining the braid. Then one may take the trace simply by adding up the linear combination of traces of basis elements. Provided n is small enough there should be no serious restriction on the length of the braid word in this process. Such an algorithm has been implemented on a computer by Morton and Short [**MS**]. In fact, the Hecke algebra method can be thought of as giving an explicit formula for the skein polynomial of the closure of an n-string braid.

Finally, another use of the Hecke algebra in knot theory is to decompose $H(q, n)$ as a direct sum of matrix algebras according to its Bratteli diagram. Ocneanu's trace will then be a weighted sum of the traces in each of the corresponding matrix algebras. The weights were first worked out by Ocneanu and are described in simple terms in [**J7**]. A proof of the formula may be found in [**We2**]. This procedure could be used to increase the number of strings accessible to a computer program calculation since the maximum storage required will only be the square of the dimension of the largest irreducible representation of the corresponding symmetric group. To the best of my knowledge no one has yet done this.

A definitive use of the weights has been made in calculating the skein polynomial of a torus knot. A torus knot is the closure of a braid $(\sigma_1 \cdots \sigma_{n-1})^m$ where n and m are relatively prime. Using the fact that $(\sigma_1 \cdots \sigma_{n-1})^n$ is in the center of the braid group and thus is represented by a scalar in any irreducible representation, one finds enough information to calculate X of a torus knot. For V this simplifies to

$$\frac{t^{(m-1)(n-1)/2}}{(1-t^2)}(1 - t^{m+1} - t^{n+1} + t^{m+n}).$$

9.4. Positivity considerations and subfactors from the Hecke algebra. Another reason for being interested in the Hecke algebra was as a vehicle for constructing subfactors of the hyperfinite II_1 factor. This was clearly understood by Ocneanu in [**F+**]. In §4.1 we constructed subfactors using the e_i algebra by showing the G.N.S. construction using the Markov trace on alg$\{1, e_1, e_2, e_3, \ldots\}$ (a C^*-algebra for $\tau^{-1} = 4\cos^2 \pi/n$ or $\tau \leq \frac{1}{4}$) and then using the closure of alg$\{1, e_2, e_3, \ldots\}$ as a subfactor, the index being τ^{-1}. We would like to do exactly the same with the Hecke algebra. The Markov trace and the Hecke algebra parameter are decoupled so the problem splits into three parts.

1) For what values of q is the Hecke algebra $H(q, \infty)$ a C^*-algebra with $*$-structure given by $\left(\frac{g_i+q}{1+q}\right)^* = \frac{g_i+q}{1+q}$?
2) Given that the answer to 1) is yes for q, for what values of z does the Markov trace define a II$_1$ factor trace on $H(q, \infty)$?
3) If the answers to 1) and 2) are yes for q and z, what is the index of the subfactor generated by $\{g_2, g_3, g_4, \ldots\}$?

There are good reasons for making the following change of variables (one should work with the elements $(g_i + q)/(1 + q)$):

$$\tau^{-1} = 2 + q + q^{-1},$$
$$\eta = \frac{z+q}{1+q}.$$

Then a considerable elaboration of Wenzl's argument in [W] shows that the answer to 1 is precisely $\tau^{-1} = 4\cos^2(\pi/n)$, $n = 3, 4, 5, \ldots$, or $0 < \tau \leq \frac{1}{4}$.

Given τ as above, the set of all values of η such that the Markov trace is positive on the C^*-algebra $H(q, \infty)$ (not the same as the algebra $H(q, \infty)$) is beautifully depicted in the following diagram due to Ocneanu. (See Figure 9.2.)

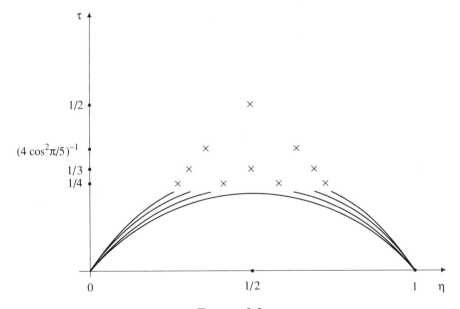

FIGURE 9.2

For all the allowed values of (τ, η) as above the trace is a II_1 factor trace. For
$$\tau^{-1} = 4\cos^2(\pi/l) \quad \text{and} \quad \eta = \frac{\sin((k+1)\pi/l)}{2\cos(\pi/l)\sin(k\pi/l)},$$
the corresponding subfactor has index $\sin^2(k\pi/l)/\sin^2(\pi/l)$ and has trivial relative commutant as shown by Wenzl in [W]. For details on all of the above see [We2].

9.5. The Birman-Murakami-Wenzl algebra. In §7.6 we defined the skein polynomial in an inductive way using the pictures L_+, L_-, L_0. The Kauffman polynomial was defined in a very similar way by considering unoriented links and the L_∞ picture. In §9.3 we saw that it can be independently derived from the Hecke algebra. Thus one is tempted to try to define an algebra having the same relation to the Kauffman polynomial as the Hecke algebra has to the skein polynomial. See [BW], [Mk].

The key to success of this project is to take seriously Kauffman's diagrammatics and consider the E_i's of §7.4 on an equal footing with braids. Thus we will try an algebra with generators $G_1, G_2, \ldots, G_{n-1}$ corresponding to braid generators, and $E_1, E_2, \ldots, E_{n-1}$ for diagrams as in §7.4. There are the obvious relations:

(BMW 1) $G_i G_j = G_j G_i$, $E_i E_j = E_j E_i$, $|i-j| \geq 2$,
(BMW 2) $G_i G_{i+1} G_i = G_{i+1} G_i G_{i+1}$,
(BMW 3) $E_i E_{i\pm 1} E_i = E_i$,
(BMW 4) $E_i^2 = k E_i$.

The algebra with this presentation is infinite-dimensional so we should add relations to cut it down to a finite-dimensional algebra, being guided by the definition of the Kauffman polynomial. The figures

suggest a linear relationship between G_i, G_i^{-1}, 1, and E_i and by symmetry it is forced to be

(BMW 5) $G_i + G_i^{-1} = x(1 + E_i)$.

This is not quite enough. Let us recall that part of the Kauffman philosophy was to replace ✕ by a ∩ . We encounter such figures when we multiply G_i's by E_i's. For instance

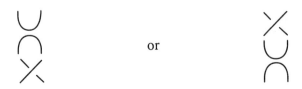

These two figures naturally give
(BMW 6) $E_i G_i = G_i E_i = x E_i$.
Also the figures

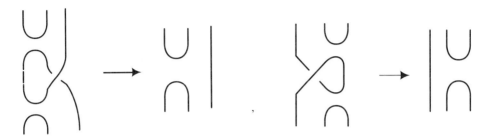

tell us to impose
(BMW 7) $E_i G_{i+1}^{\pm 1} E_i = a^{\mp 1} E_i$, $\quad E_i G_{i-1}^{\pm 1} E_i = a^{\mp 1} E_i$.
We leave the reader to justify
(BMW 8) $E_i G_{i\pm 1} G_i = E_i E_{i\pm 1}$.

It is not too hard to see that the algebra presented by (BMW 1) → (BMW 8) is finite-dimensional. Indeed, consider all the $1 \cdot 3 \cdot 5 \cdots (2n-1)$ ways of partitioning a set with $2n$ elements into subsets of size two. Split the $2n$ elements into two subsets of size n and display them on two parallel lines. For each partition join up the pairs of points with curves between the parallel lines. The curves joining two points on the bottom row should not intersect the (equally many) curves joining points on the top row. It is also convenient if these curves are nested so as not to intersect each other at all. For instance for $n = 6$, let the partition be $\{1, 2\}, \{3, 5\}, \{4, 6\}$. Then we would obtain the picture

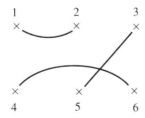

In general, there will be crossings between the joining curves. Convert these crossings into over or undercrossings at will (of course a good convention would be useful at this point, but not necessary for our discussion). Thus we might end up with the diagram

from our example.

It is now a long exercise to see that all the diagrams so obtained can be written as words on the BMW generators and that the obvious multiplication by concatenation produces a linearly closed system using (BMW 1) → (BMW 8). Thus the algebra presented by these relations is of dimension $\leq 1 \cdot 3 \cdot 5 \cdots (2n-1)$.

Note that (BMW 5) and (BMW 6) imply

$$a + a^{-1} = x(1+k) \text{ or } k = (a + a^{-1} - x)/x.$$

We define the BMW algebra, with parameters a and x, to be the algebra $\mathrm{BMW}(a, x, n)$ presented on generators $G_1, G_2, \ldots, G_{n-1}, E_1, E_2, \ldots, E_{n-1}$, with relations (BMW 1) → (BMW 8). These relations are not supposed to be a minimal set.

9.6. The Markov trace on the BMW algebra. So far we have defined an algebra generated by braids with only a superficial look at the defining philosophy of the Kauffman polynomial. But remember that there should also be a trace on the algebra such that the trace of a braid is, up to normalization, the Kauffman polynomial of the closed braid. In particular, we must have special behaviour under the type II Markov move. It is shown in [**BW**] how one might prove the existence of a trace on $\bigcup_{n=1}^{\infty} \mathrm{BMW}(z, x, n)$ uniquely defined by the following axioms:

$$\left. \begin{array}{l} \mathrm{tr}(1) = 1, \\ \mathrm{tr}(AB) = \mathrm{tr}(BA), \\ \mathrm{tr}(AG_n) = \dfrac{a^{-1}x}{a + a^{-1} - x} \mathrm{tr}(A), \\ \mathrm{tr}(AE_n) = \dfrac{x}{a + a^{-1} - x} \mathrm{tr}(A). \end{array} \right\} \text{ for } A \in \mathrm{BMW}(z, x, n).$$

It is not surprising (though still not trivial) that if $\alpha \in B_n$ is a braid and $\pi \colon B_n \to \mathrm{BMW}(a, x, n)$ is the obvious representation $\sigma_i \mapsto G_i$, then

$$F_{\hat{\alpha}}(x, a) = \left(\frac{a + a^{-1} - x}{x} \right)^{n-1} a^{e(\alpha)} \mathrm{tr}(\pi(\alpha))$$

(e is exponent sum as in §7.1).

The reason this formula is not obvious is that closed braids are oriented, so it is a little tricky to handle the "L_∞" picture. In the approach of [**BW**] this problem does not occur since one defines tr using the existence of the Kauffman polynomial! An approach more in line with the Hecke algebra approach to P_L would be to prove that the $2^{-n} n! \binom{2n}{n}$ elements defined in §9.4 are a basis of $\mathrm{BMW}(a, x, n)$ and that the above axioms define a trace.

One could then use the plat approach of §7.3 to show the relationship with F_L. To the best of our knowledge this program has not been carried out.*

9.7. Structure of the BMW algebra.
One would like to know the following:
1) For what values of (a, x, n) is BMW(a, x, n) semisimple?
2) For the values as in 1, what is the Bratteli diagram for BMW?
3) What are the weights of the Markov trace?

I believe that all these three are known to Wenzl, but not yet published;** so let me give the philosophy of his approach, which will enable us to obtain the "generic" Bratteli diagram.

The idea is to exploit a purely algebraic version of the basic construction of §2.6 and §2.7. The detailed use will depend on the nondegeneracy of the Markov trace, which can be decided by working out the weights inductively. It is clear, however, that only rational functions of a and x will appear so that for almost all (a, x) there is no problem and one may proceed as follows. Note that the BMW and trace axioms imply the following:

$$E_n \operatorname{BMW}(a, x, n) E_n = \operatorname{BMW}(a, x, n-1) E_n$$

and $\operatorname{tr}(AE_n) = \operatorname{tr}(A)\operatorname{tr}(E_n)$ for $A \in \operatorname{BMW}(a, n)$. So we know that BMW$(a, x, n+1)$ contains, as a two-sided ideal, a copy of the basic construction for the pair BMW$(a, x, n-1) \subset$ BMW(a, x, n). More use of the BMW relations shows inductively that this ideal is exactly the two-sided ideal generated by E_n. Moreover, if we put $E_n = 0$ then the BMW presentation becomes precisely the Hecke algebra presentation (with $g_i = \sqrt{-q}G_i$, $x = \frac{q-1}{\sqrt{-q}}$) so that by semisimplicity the ideal above has a complementary ideal isomorphic to the Hecke algebra.

Thus we can now draw the Bratteli diagram, noting that the Hecke algebra parts must give the Hecke algebra as a sub-diagram. There is a "reflected part" and a Hecke algebra part. Identifying the reflected parts with the Young diagrams that gave rise to them we get Figure 9.3, which appears in [**BW**].

Note how the sums of the squares add up to $1 \cdot 3 \cdot 5 \cdots (2n-1)$. This was proved from the combinatorics of the diagram by Kidwell ($225 + 100 + 100 + 400 + 120 = 945 = 1 \times 3 \times 5 \times 7 \times 9$). We emphasize again that this is only the picture for generic a and x. The situation for special values is understood by Wenzl who has analyzed the positivity of the trace and possible C^*-algebra structures and has thus produced many interesting new subfactors (see §9.4 for the simpler situation of the Hecke algebra) in [**We3**].

Also note the existence, for $n = 2$, of an interesting two-parameter family of representations of B_3. Explicit matrices representing σ_1 and σ_2 in this representation may be found in [**BW**].

---
*See a forthcoming paper by Morton and Wassermann.
See [We3**].

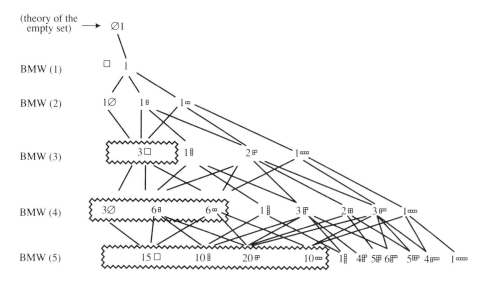

FIGURE 9.3

9.8. Wenzl's result on Brauer's centralizer algebra. A basic problem of invariant theory is to take a representation of a group G on a vector space V and decompose the tensor powers of the representation. This is equivalent, in the semisimple case, to decomposing the algebra, sometimes called the centralizer algebra, of endomorphisms of $\bigotimes^n V$ that commute with the action of G. The case $G = \mathrm{GL}(V)$ has long been understood. It is obvious that the action of the symmetric group S_n on $\bigotimes^n V$ (coming from permuting the tensor product components) commutes with the action of $\mathrm{GL}(V)$. The fact that the centralizer of $\mathrm{GL}(V)$ on $\bigotimes^n V$ is the algebra generated by the symmetric group was probably known to Frobenius.

For groups smaller than $\mathrm{GL}(V)$ we expect the centralizer algebras to be larger. Indeed, if G preserves a nondegenerate bilinear form $\langle , \rangle$, there is a simple way to create invariants. For V is isomorphic to V^* via $\langle , \rangle$, so the canonical element of $V \otimes V^*$ can be viewed as an element of $V \otimes V$ which will clearly define a one-dimensional invariant subspace of $V \otimes V$. Orthogonal projection onto this subspace will be an operator e in $\mathrm{End}(V \otimes V)$ commuting with the action of G. (One should be a bit more careful if $\langle , \rangle$ is not symmetric.) Using the notation of the appendix one obtains elements $e_1, e_2, \ldots, e_{n-1}$ in $\mathrm{End}(\bigotimes^n V)$ which are in the centralizer of G. It is easy to check that $e_i e_j = e_j e_i$ for $|i - j| \geq 2$ and $e_i e_{i \pm 1} e_i = (\dim V)^{-1} e_i$. Using the generators of the symmetric group as G_i's we find an algebraic system very similar to the BMW algebra of §9.6. Indeed Brauer in [**Br**] defined an abstract algebra whose basis is given by the partitions of a set of size $2n$ into subsets of size 2 as we described in §9.6, but forgetting all crossing data. The multiplication law in the algebra is defined by concatenation as in §9.6,

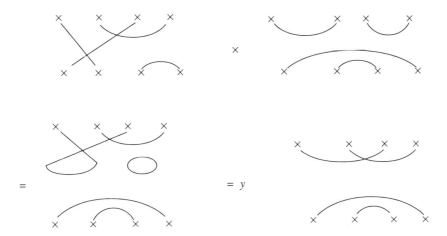

FIGURE 9.4

only every time one obtains a closed loop one eliminates it and multiplies by a power of a parameter y. (See Figure 9.4.)

One may verify directly that this multiplication law is associative, so one obtains an algebra $D_n(y)$ for each complex number y. Putting $y = \dim V$ we get a map from $D_n(y)$ to the centralizer of the orthogonal group on $\bigotimes^n V$ by sending G_i to the obvious symmetric group element and e_i to $(\dim V)e_i$. This map is *onto* the centralizer of the orthogonal group and appropriate modifications work for the symplectic group. Hanlon and Wales studied the algebra $D_n(y)$ when y is not an integer and conjectured that it is semisimple, see [**HW**].

In [**We4**], Wenzl used the similarity of the Brauer algebra to the BMW algebra to prove the conjecture and give the structure of the centralizer algebra completely in terms of the Bratteli diagram of §9.6. There is a subtlety caused by the fact that the Brauer algebra is a limit rather than a specialization of the BMW algebra as we have defined it, so Wenzl adapted the techniques used to determine the structure of the BMW algebra to the simpler case of the Brauer algebra. This must be considered the most spectacular success of the BMW algebra and the basic construction technique.

9.9. Quantum invariant theory. In §8.6 we described rather briefly the notion of quantum group. We said that the enveloping algebra of any complex simple Lie algebra may be deformed to become a noncommutative, noncocommutative Hopf algebra $U_h(\mathcal{G})$ with a privileged element R in $U_h(\mathcal{G}) \otimes U_h(\mathcal{G})$, and that all the finite-dimensional representations deform also. In the spirit of §9.7 we may consider "quantum invariant theory" to be the decomposition into irreducibles of the tensor powers of a representation. Jimbo showed in [**Ji**] that the duality between $GL(k)$ and the symmetric groups also deforms into a duality between $U_h(\mathrm{gl}(k))$ (or $U_h(\mathrm{sl}_k)$) and the

Hecke algebra—the transpositions acting on $\bigotimes^n V (\dim V = k)$ being replaced by the braid generators R_i obtained from the R matrix as in §5.7. It follows immediately from the Yang-Baxter equation that the R_i's all belong to the commutant of the tensor powers of the fundamental representation, and Jimbo observed that they generate this commutant (at least for generic values of the deformation parameter).

More or less the same is true for the B_n, C_n, and D_n series of Lie algebras, in their defining representations. The symmetric group is replaced by the Brauer algebra and the Hecke algebra by the BMW algebra. But notice a remarkable thing that happens here. In the classical case, the symmetric group does not generate the centralizer, whereas in the quantum case, the braid group generators, having three distinct eigenvalues, do (again for generic values)! Thus, the extra invariants needed to obtain the whole centralizer appear simply as degenerations of the quantum situation.

In [KR] it is shown that the same is true for *all* finite-dimensional irreducible representations V of sl_2—the braid generators generate the centralizer of $U_n(sl_2)$ in $\bigotimes^n V$.

One cannot hope for this to be always true, since for many representations (e.g., the adjoint representation of sl_3), the centralizer on $V \otimes V$ is noncommutative while the braid group B_2 is commutative. However, one has not yet used the spectral parameter. There is no reason why the $\check{R}(\lambda)$'s should commute for all values of the spectral parameters, but they are certainly in the centralizer. One is tempted to conjecture that for any irreducible representation the centralizer is generated by the $\check{R}(\lambda)$'s, though I have no powerful evidence for this. There seem to be many interesting questions in this area.

Appendix

A notation for algebras and endomorphisms. The following notation has been frequently used. If A is an associative algebra with identity 1, we shall define many embeddings of $A \otimes A$ in $\bigotimes^n A = A \otimes A \otimes \cdots \otimes A$ (n copies of A). First observe that the symmetric group S_n acts on $\bigotimes^n A$ by algebra automorphisms in the obvious way. If $\sigma \in S_n$ is a permutation let $\pi(\sigma)$ be the corresponding automorphism. If $R \in A \otimes A$ we let $R_{12} \in \bigotimes^n A$ be $R \otimes 1 \otimes 1 \otimes \cdots \otimes 1$ ($n-2$ 1's). Then for $1 \leq i \neq j \leq n$ we define R_{ij} to be $\pi(\sigma)(R_{12})$ where σ is any permutation sending 1 to i and 2 to j. If $R = \sum_m a_m \otimes b_m$ then clearly

$$R_{ij} = \sum_m 1 \otimes \cdots \otimes a_m \otimes \cdots \otimes b_m \otimes \cdots \otimes 1$$

$$\qquad\qquad\qquad\uparrow \qquad\quad \uparrow$$
$$\qquad\qquad\quad i\text{th slot} \quad j\text{th slot}$$

The element R_{ii+1} will be denoted R_i.

In the special case $A = \text{End}(V)$ we use also the canonical identification of $\bigotimes^n A$ with $\text{End}(\bigotimes^n V)$ to produce elements $R_{ij} \in \text{End}(\bigotimes^n V)$ from $R \in \text{End}(V \otimes V)$.

Also, if B is an algebra and A is $\text{End}(V)$ as above then given an element $T \in B \otimes A$ (a $\dim V \times \dim V$ matrix of elements of B) we use the notation T_1 for $(T \otimes 1 \otimes \cdots \otimes 1)$ in $B \otimes A \otimes (\bigotimes^{n-1} A)$ and T_i for the element $\pi(\sigma)$ applied to T_1 where σ is a permutation exchanging 1 and n and S_n acts in the obvious way on $B \otimes (\bigotimes^n A)$. This is in conflict with our previous notation if $B \subset A$.

References

[ADW] Y. Akutsu, T. Deguchi, and M. Wadati, *The Yang-Baxter relation: A new tool for knot theory*, in *Braid Groups, Knot Theory and Statistical Mechanics*, ed. C. Yang, M. Ge. World Scientific (1989), 151–200.

[Ax1] J. Alexander, *Topological invariants of knots and links*, Trans. Amer. Math. Soc. **20** (1923), 275–306.

[Ax2] ___, *A lemma on systems of knotted curves*, Proc. Nat. Acad. **9** (1923), 93–95.

[ABF] G. Andrews, R. Baxter, and P. Forrester, *Eight-vertex SOS model and generalized Rogers-Ramanujan type identities*, J. Stat. Phys. **35** (1984), 193–266.

[AW] Y. Akutsu and M. Wadati, *Knot invariants and the critical statistical systems*, J. Phys. Soc. Japan **56** (1987), 839–842.

[AM] G. Anderson and G. Moore, *Rationality in conformal field theory*, Preprint, 1988.

[Ar] E. Artin, *Theory of braids*, Ann. Math. **48** (1947), 101–126.

[Ba] R. Baxter, *Exactly Solved Models in Statistical Mechanics*, Academic Press, London, 1982.

[BPZ] A. Belavin, A. Polyakov, and A. Zamolodchikov, *Infinite conformal symmetries in two-dimensional quantum field theory*, Nucl. Phys. B **241** (1984), 333–380.

[Bi1] J. Birman, *Braids, links, and mapping class groups*, Ann. Math. Studies **82**, Princeton University Press, 1976.

[Bi2] ___, *On the stable equivalence of plat representations of knots and links*, Canad. J. Math. **28** (1976), 264–290.

[BW] J. Birman and H. Wenzl, *Braids, link polynomials and a new algebra*, Trans. Amer. Math. Soc. **313** (1989), 249–273.

[Bo] N. Bourbaki, *Groupes et Algèbres de Lie*, IV, V, et VI, Hermann, Paris, 1968.

[BLM] R. Brandt, W. Lickorish, and K. Millett, *A polynomial invariant for unoriented knots and links*, Invent. Math. **84** (1986), 563–573.

[Br] R. Brauer, *On algebras which are connected with the semisimple continuous groups*, Ann. Math. **38** (1937), 854–887.

[BZ] G. Burde and H. Zieschang, *Knots*, de Gruyter Studies in Math. **5**, Walter de Gruyter & Co., 1985.

[CIZ] A. Cappelli, C. Itzykson, and J.-B. Zuber, *Modular invariant partition functions in two dimensions*, Nucl. Phys. B **280** (1987), 445–465.

[Ch] E. Christensen, *Subalgebras of a finite algebra*, Math. Ann. **243** (1979), 17–29.

[Co1] A. Connes, *Une classification des facteurs de type* III, Ann. Sci. Ec. Norm. Sup. **6** (1973), 133–252.

[Co2] ———, *A factor of type* II_1 *with countable fundamental group*, J. Operator Theory **4** (1980), 151–153.

[Co3] ———, *Periodic automorphisms of the hyperfinite factor of type* II_1, Acta Sci. Math. **39** (1971), 39–66.

[Co4] ———, *Outer conjugacy classes of automorphisms of factors*, Ann. Sci. Ec. Norm. Sup. **8** (1975), 383–419.

[Co5] ———, *Sur la classification des facteurs de type* II, C. R. Acad. Sci. Paris **281** (1975), 13–15.

[Co6] A. Connes, *Classification of injective factors*, Ann. Math. **104** (1976), 73–115.

[CJ] A. Connes and V. Jones, *Property T for von Neumann algebras*, Bull. Lond. Math. Soc. **17** (1985), 57–62.

[CS] A. Connes and E. Størmer, *Entropy of automorphisms of* II_1 *von Neumann algebras*, Acta Math. **134** (1975), 289–306.

[CT] A. Connes and M. Takesaki, *The flow of weights on a factor of type* III, Tohoku Math. J. **29** (1977).

[Co] J. Conway, *An enumeration of knots and links*, in *Computational Problems in Abstract Algebra*, ed. J. Leech, Pergamon Press (1969), 329–358.

[Di1] ———, *Les algèbres d'opérateurs dans l'espace Hilbertien*, Gauthier-Villars, Paris, 1957.

[Di2] J. Dixmier, *Formes linéaires sur un anneau d'opérateurs*, Bull. Soc. Math. France **81** (1953), 9–39.

[Dr] V. Drinfeld, *Quantum groups*, in Proc. ICM 86, Vol. 1, 798–820.

[Fa] L. Faddeev, *Integrable models in* $1+1$-*dimensional quantum field theory*, in Les Houches Lectures 1982, Elsevier, Amsterdam (1984), 563–608.

[FM] J. Feldman and C. Moore, *Ergodic equivalence relations, cohomology and von Neumann algebras* I, II, Trans. Amer. Math. Soc. **234** (1977), 289–324, 325–361.

[FW] J. Franks and R. Williams, *Braids and the Jones-Conway polynomial*, Trans. Amer. Math. Soc. **303** (1987), 97–108.

REFERENCES

[Ft] P. Freyd, D. Yetter, J. Hoste, W. Lickorish, K. Millett, A. Ocneanu, *A new polynomial invariant of knots and links*, Bull. Amer. Math. Soc. **12** (1985), 183–190.

[FQS] D. Friedan, Z. Qiu, and S. Shenker, *Conformal invariance, unitarity and two-dimensional critical exponents*, in *Vertex Operators in Mathematics and Physics*, Ed. J. Lepowsky, S. Mandelstam, and I. Singer, Springer-Verlag, MSRI pub. **3** (1984), 419–450.

[Ga] F. Garside, *The braid group and other groups*, Quart. J. Math. Oxford **20** (1969), 235–254.

[GKO] P. Goddard, A. Kent, and D. Olive, *Unitary representations of the Virasoro and Super-Virasoro algebras*, Comm. Math. Phys. **103** (1986), 105–119.

[Go] M. Goldman, *On subfactors of factors of type* II_1, Michigan Math. J. **7** (1960), 167–172.

[GJ] D. Goldschmidt and V. Jones, *Metaplectic link invariants*, Geometriae Dedicata **31** (1989), 165–191.

[Gl] D. Goldschmidt, *Notes on symmetric groups and Hecke algebras*, Berkeley (1989).

[GHJ] F. Goodman, P. de la Harpe, and V. Jones, *Coxeter graphs and towers of algebras*, MSRI Publications (Springer) **14** (1989).

[Ha] U. Haagerup, *Connes' bicentralizer problem and the uniqueness of the injective factor of type* III_1, Acta Math. **158** (1987), 95–148.

[HW] P. Hanlon and D. Wales, *On the decomposition of Brauer's centralizer algebras*, J. Algebra **121** (1989), 409–445.

[HKW] P. de la Harpe, M. Kervaire, and C. Weber, *On the Jones polynomial*, L'Enseignement Mathématique **32** (1986), 271–335.

[He] E. Hecke, *Über die Bestimmung Dirichletscher Reihen durch ihre Funktionalgleichung*, Math. Annalen **112** (1936), 664–699.

[Ho] C. Ho, *A new polynomial for knots and links—preliminary report*, Abstracts Amer. Math. Soc. **6**, 4 (1985), 300. Abstract 821-57-16.

[Hop] E. Hopf, *Theory of measures and invariant integrals*, Trans. Amer. Math. Soc. **34** (1932), 373–393.

[HLS] R. Hoegh-Krohn, M. Landstad, and E. Størmer, *Compact ergodic groups of automorphisms*, Ann. Math. **114** (1981), 75–86.

[Ji1] M. Jimbo, *Quantum R-matrix for the generalized Toda system*, Comm. Math. Phys. **102** (1986), 537–547.

[Ji2] M. Jimbo, *A q-difference analogue of* $U(\mathfrak{G})$ *and the Yang-Baxter equation*, Lett. Math. Phys. **10** (1986), 63–69.

[J1] V. Jones, *An invariant for group actions*, in *Algèbres d'opérateurs*, Springer Lect. Notes **725** (1978), 237–253.

[J2] V. Jones, *Actions of Finite Groups on the Hyperfinite Type* II_1 *Factor*, Memoirs Amer. Math. Soc. **237** (1980).

[J3] V. Jones, *Index for subfactors*, Invent. Math. **72** (1983), 1–25.

[J4] V. Jones, *Braid groups, Hecke algebras and type* II_1 *factors*, in *Geometric Methods in Operator Algebras*, ed. Araki and Effres, Pitman Res. Notes in Math. (1983), 242–273.

[J5] V. Jones, *On knot invariants related to some statistical mechanical models*, Pacific J. Math. **137** (1989), 311–334.

[J6] V. Jones, *A polynomial invariant for knots via von Neumann algebras*, Bull. Amer. Math. Soc. **12** (1985), 103–111.

[J7] V. Jones, *Hecke algebra representations of braid groups and link polynomials*, Ann. Math. **126** (1987), 335–388.

[J8] V. Jones, Letter to L. Kauffman, Nov. 1986.

[JT] V. Jones and M. Takesaki, *Actions of compact abelian groups on semifinite injective factors*, Acta Math. **153** (1984), 213–258.

[Ka1] V. Kac, *Contravariant form for infinite-dimensional Lie algebras and superalgebras*, Springer Lecture Notes in Physics **94** (1979), 441–445.

[Ka2] V. Kac, *Infinite-dimensional Lie Algebras*, Boston, Birkhäuser, 1983.

[K1] L. Kauffman, *State models and the Jones polynomial*, Topology **26** (1987), 395–401.

[K2] L. Kauffman, *An invariant of regular isotopy*, Trans. Amer. Math. Soc. **318** (1990), 417–471.

[K3] L. Kauffman, *New invariants in the theory of knots*, Amer. Math. Monthly **95** (1988), 195–242.

[KR] A. Kirillov and N. Reshetikhin, *Representations of the algebra* $U_q(sl(2))$, *q-orthogonal polynomials, and invariants of links*, LOMI preprint E-9-88, Steklov Math. Inst., Leningrad.

[Koh1] T. Kohno, *Linear representations of the braid groups and the classical Yang-Baxter equations*, Ann. Inst. Fourier **37**, 4 (1987), 139–160.

[Koh2] T. Kohno, *Topological invariants for 3-manifolds using representations of mapping class groups* I, Preprint, Kyushu University, 1990.

[Kos] H. Kosaki, *Extension of Jones' theory on index to arbitrary factors*, J. Funct. Analysis **66** (1986), 123–140.

[Kos1] H. Kosaki, *AFD factor of type* III_0 *with many isomorphic index 3 subfactors*, University of Colorado, Boulder, preprint, 1991.

[Kr] W. Krieger, *On ergodic flows and the isomorphism of factors*, Math. Annalen **223** (1976), 19–70.

[Li] E. Lieb, Phys. Rev. **162** (1967), 162–172.

[Lo] P. Loi, *On the theory of index and type* III *factors*, Pennsylvania State University thesis, 1988.

[Lu] G. Lusztig, *Quantum deformations of certain simple modules over enveloping algebras*, Adv. Math. **70** (1988), 237–249.

[McD] D. McDuff, *Central sequences and the hyperfinite factor*, Proc. Lond. Math. Soc. **21** (1970), 443–461.

[Ma] A. Markov, *Über die freie Aquivalenz geschlossener Zöpfe*, Math. Sb. **1**

(1935), 73–78.

[Mi] J. Milnor, *On the total curvature of knots*, Ann. Math. **52** (1950), 248–257.

[MS] G. Moore and N. Seiberg, *Classical and quantum conformal field theory*, Comm. Math. Phys. **123** (1989), 177–254.

[Mor1] H. Morton, *Threading knot diagrams*, Math. Proc. Cambridge Philos. Soc. **99** (1986), 247–260.

[Mor2] H. Morton, *Closed braid representations for a link and its 2-variable polynomial*, preprint, Univ. of Liverpool, 1986.

[Mos] G. Mostow, *Quasi-conformal mappings in n-space and the rigidity of hyperbolic space forms*, Publ. Math. IHES **34** (1967), 53–104.

[MvN1] F. Murray and J. von Neumann, *On rings of operators*, Ann. Math. **37** (1936), 116–229.

[Mvn2] F. Murray and J. von Neumann, *On rings of operators*, II, Trans. Amer. Math. Soc. **41** (1937), 208–248.

[MvN3] F. Murray and J. von Neumann, *On rings of operators*, IV, Ann. Math. **44** (1943), 716–808.

[Mk] J. Murakami, *The Kauffman polynomial of links and representation theory*, Osaka J. Math. **24** (1987), 745–758.

[Ms] K. Murasugi, *Jones' polynomials and classical conjectures in knot theory*, Topology **26** (1987), 187–194.

[NT] N. Nakamura and Z. Takeda, *On the extensions of finite factors*, I, II, Proc. Jap. Acad. **35** (1959), 149–156 and 215–220.

[N] E. Nelson, *Non-commutative integration theory*, J. Funct. Analysis **15** (1974), 103–116.

[Oc1] A. Ocneanu, *Actions of discrete amenable groups on von Neumann algebras*, Springer Lect. Notes in Math. **1138** (1985).

[Oc2] A. Ocneanu, *Quantized groups, string algebras and Galois theory for algebras*, in *Operator Algebras and Applications*, ed. Evans and Takesaki (1988), 119–172.

[On] L. Onsager, *Crystal statistics I, A two-dimensional model with an order-disorder transition*, Physics Reports **65** (1944), 117–149.

[Pa] V. Pasquier, *Etiology of IRF models*, Comm. Math. Phys. **118** (1988), 335–364.

[Ph] J. Phillips, *Automorphisms of full II_1 factors with applications to factors of type III*, Duke Math. J. **43** (1976), 375–385.

[PP1] M. Pimsner and S. Popa, *Entropy and index for subfactors*, Ann. Sci. Ec. Norm. Sup. **19** (1986), 57–106.

[PP2] M. Pimsner and S. Popa, *Sur les sous-facteurs d'indice fini d'un facteur de type II_1 ayant la propriété T*, C. R. Acad. Sci. Paris **303** (1986), 359–361.

[Pop] S. Popa, *Classification of subfactors: Reduction to commuting squares*, Invent. Math. **101** (1990), 19–43.

[Pow] R. Powers, *Representations of uniformly hyperfinite algebras and their associated von Neumann rings*, Ann. Math. **86** (1967), 138–171.

[PS] A. Pressley and G. Segal, *Loop Groups*, Oxford Math. Monographs, 1986.

[PT] J. Przytycki and P. Traczyk, *Invariants of links of Conway type*, Kobe J. Math. **4** (1987), 115–139.

[Rei] K. Reidemeister, *Knotentheorie*, Ergebn. Math. Grezgeb. (Springer) **1** (1932).

[RS] K. Reidemeister, *Zür dreidimensionalen topologie*, Abh. Math. Sem. Univ. Hamburg **9** (1933), 189–194.

[Res] N. Reshetikhin, *Quantized universal enveloping algebras, the Yang-Baxter equation and invariants of links* I,II, LOMI Preprints, Leningrad, 1988.

[Rob] A. Robert, *Introduction to the Representation Theory of Compact and Locally Compact Groups*, Cambridge Univ. Press, 1983.

[Rol] D. Rolfsen, *Knots and Links*, Publish or Perish, 1976.

[Ros1] M. Rosso, *Groupes quantiques et modèles à vertex de V. Jones en théorie des noeuds*, C. R. Acad. Sci. Paris **307** (1988), 207–210.

[Ros2] M. Rosso, *Finite-dimensional representations of the quantum analogue of the enveloping algebra of a complex simple Lie algebra*, Comm. Math. Phys. **117** (1988), 581–593.

[Sa] S. Sakai, C^*-*algebras and* W^*-*algebras*, Springer-Verlag, 1983.

[Sei] H. Seifert, *Über das Geschlecht von Knoten*, Math. Ann. **110** (1934), 571–592.

[Ser] J. P. Serre, *A Course in Arithmetic*, Springer G.T.M. series, 1973.

[ST] G. Shephard and J. Todd, *Finite unitary reflection groups*, Canad. J. Math. **6** (1954), 274–304.

[Sk] C. Skau, *Finite subalgebras of a von Neumann algebra*, J. Funct. Anal. **25** (1977), 211–235.

[Sq] C. Squier, *The Burau representation is unitary*, Preprint, 1983.

[Sun] V. Sunder, *A model for AF algebras and a representation of the Jones projections*, J. Operator Theory **18** (1987), 289–301.

[Suz] N. Suzuki, *Crossed products of rings of operators*, Tohoku Math. J. **11** (1959), 113–124.

[Ta1] M. Takesaki, *Tomita's theory of modular Hilbert algebras and its applications*, Springer Lect. Notes in Math. **128** (1970).

[Ta2] M. Takesaki, *Duality for crossed products and the structure of von Neumann algebras of type* III, Acta Math. **131** (1973), 249–310.

[TL] N. Temperley and E. Lieb, *Relations between the "percolation" and "colouring" problem and other graph-theoretical problems associated with regular planar lattices: some exact results for the "percolation" problem*, Proc. Royal Soc. Ser. A **322** (1971), 251–280.

[Th] M. Thistlethwaite, *A spanning tree expansion of the Jones polynomial*,

Topology **26** (1987), 297–309.

[Tr] B. Trace, *On the Reidemeister moves of a classical knot*, Proc. Amer. Math. Soc. **89** (1983), 722–724.

[TK] A. Tsuchiya and Y. Kanie, *Vertex operators in conformal field theory on $\mathbb{P}^1$ and monodromy representations of braid groups*, Adv. Stud. Pure Math. **16** (1988), 297–372.

[Tu1] V. Turaev, *A simple proof of the Murasugi and Kauffman theorems on alternating links*, Enseign. Math. **33** (1987), 203–225.

[Tu2] V. Turaev, *The Yang-Baxter equation and invariants of links*, Invent. Math. **92** (1988), 527–553.

[U] H. Umegaki, *Conditional expectation in an operator algebra*, I, Tohoku Math. J. **6** (1954), 177–181.

[vN1] J. von Neumann, *Zür algebra der funktionaloperatoren*, Math. Ann. **102** (1929), 370–427.

[vN2] J. von Neumann, *On rings of operators*, III, Ann. Math. **37** (1936), 111–115.

[Wa] A. Wassermann, *Ergodic actions of compact groups on operator algebras*, III: *Classification for* $SU(2)$, Invent. Math. **93** (1988), 309–355.

[We1] H. Wenzl, *On sequences of projections*, C. R. Math. Rep. Acad. Sci. Canada **9** (1987), 5–9.

[We2] H. Wenzl, *Hecke algebras of type A_n and subfactors*, Invent. Math. **92** (1988), 349–383.

[We3] H. Wenzl, *Quantum groups and subfactors of type B, C, and D*, Comm. Math. Phys. **133** (1990), 383–432.

[We4] H. Wenzl, *On the structure of Brauer's centralizer algebras*, Ann. Math. **128** (1988), 173–193.

[Wi] E. Witten, *Quantum field theory and the Jones polynomial*, Comm. Math. Physics **121** (1989), 351–399.

[Wo] S. Woronowicz, *Compact matrix pseudogroups*, Comm. Math. Phys. **111** (1987), 613–665.

[Y] S. Yamada, *The minimum number of Seifert circles equals the braid index of a link*, Invent. Math. **89** (1987), 347–356.